Green Engineering and Sustainable Technology

Scrivener Publishing
100 Cummings Center, Suite 541J
Beverly, MA 01915-6106

Publishers at Scrivener
Martin Scrivener (martin@scrivenerpublishing.com)
Phillip Carmical (pcarmical@scrivenerpublishing.com)

Green Engineering and Sustainable Technology

Tools and Techniques

Edited by

Senthilkumar Rathnasamy

School of Chemical and Biotechnology, SASTRA Deemed to be University, Thanjavur, India

Vivek Rangarajan

Department of Chemical Engineering, Birla Institute of Technology and Science Pilani KK Birla Goa Campus, Sancoale, India

and

Umile Gianfranco Spizzirri

Ionian Department of Law, Economics and Environment, University of Bari Aldo Moro, Taranto, Italy

Scrivener Publishing

WILEY

This edition first published 2026 by John Wiley & Sons, Inc., 111 River Street, Hoboken, NJ 07030, USA and Scrivener Publishing LLC, 100 Cummings Center, Suite 541J, Beverly, MA 01915, USA

For more information about Scrivener publications please visit www.scrivenerpublishing.com.

Wiley Global Headquarters
111 River Street, Hoboken, NJ 07030, USA

For details of our global editorial offices, customer services, and more information about Wiley products visit us at www.wiley.com.

The manufacturer's authorized representative according to the EU General Product Safety Regulation is Wiley-VCH GmbH, Boschstr. 12, 69469 Weinheim, Germany, e-mail: Product_Safety@wiley.com.

Library of Congress Cataloging-in-Publication Data

ISBN 978-1-394-30340-3

Cover image: Generated with AI using Adobe Firefly
Cover design by Russell Richardson

Set in size of 11pt and Minion Pro by Manila Typesetting Company, Makati, Philippines

Printed in the USA

10 9 8 7 6 5 4 3 2 1

Contents

Preface

"***Sustainability is no longer about doing less harm. It's about doing more good.***" - *Jochen Zeitz*

"***Progress is impossible without change, and those who cannot change their minds cannot change anything.***" - *George Bernard Shaw*

The growing rate of industrialization fosters environmental challenges, prompting them to think of greener, safer, and more sustainable alternatives. "Green Engineering and Sustainable Technology" focuses on green engineering practices that provide innovative solutions to existing problems and open up new processes and products that uphold sustainability.

The main goal is to bridge the gap between green engineering concepts and practical applications by providing insights into how green principles developed into sustainable technology in industries to promote a circular economy. Following green metrics along with technological advancements that measure sustainability helps to reduce the ecological footprint. These metrics help the organization assess the environmental impacts of its process in terms of energy consumption, carbon emissions, and waste generation to make strategic decisions for eco-friendly manufacturing.

The green solvents proposed here play a major role in green engineering and sustainable technology in reducing environmental impact with safety improvement and increased process efficiency. Solvents such as ionic liquids, deep eutectic solvents, and terpenes minimize the emissions with tunable solvation and improve reaction efficiency. Another chapter revealed the need for process intensification to enhance production efficiency through the synergetic effects of integration addressing environmental regulations and sustainability goals. All the integrated techniques in the book delve into groundbreaking developments in environmentally and economically advantageous sustainable technology. Nanoemulsion addressed in the chapter discusses the advantages of emulsification in the food and drug industry and its sustainability in industrial applications.

In continuation, techno-economic analysis is addressed here for the economic feasibility, investment decision-making, technology optimization for scale-up, and the green technology advancement for commercialization of any product in the market analysis. In addition, life cycle analysis is needed for waste management, environmental evaluation, uncertainty, sensitivity analysis, and hotspot identification. Both these chapters have a critical role in green engineering and sustainability technology development in providing a framework for supporting the transition to a green economy and attaining sustainability goals.

Industrial sustainable fermentation for acids, enzymes, and secondary metabolite production focuses on waste minimization for environmental sustainability. Sustainable fermentation utilizing renewable sources shifts industries toward a more sustainable production of biofuels, biochemicals, etc. Moreover, green electrolytes are designed to enhance the safety of engines, particularly in the upcoming field of electric vehicles reduce pollution during production, usage, and disposal. This sustainable energy system supports fuel economy and increases the demand for green electrolytes in the electronic field and electric vehicle manufacturing.

We hope that this book will inspire *engineers*, *researchers*, and *students* to apply green engineering principles in various fields such as biotechnology, chemical, civil, computer, environmental, and material science, who think beyond conventional practices and seek technologically advanced solutions. We would like to thank Wiley and Scrivener Publishing for their continuous support and guidance in the production of this volume.

Senthilkumar Rathnasamy
SASTRA University, India
Vivek Rangarajan
BITS Pilani Goa Campus, India
Umile Gianfranco Spizzirri
Ionian Department of Law, Economics and Environment,
University of Bari Aldo Moro, Taranto, Italy

1

Introduction to Green Engineering: Principles, Approach, Applications, and Sustainability

Neela Gayathri Ganesan[1*], Rami A. Abdel-Rahem[2], Vivek Rangarajan[3†], Sivalingam Ramesh[4], Cédric Delattre[5] and K. Senthilkumar[6]

[1]*Department of Chemical and Bio-Chemical Engineering, Dongguk University, Seoul, South Korea*
[2]*Department of Chemistry, Faculty of Arts and Sciences, University of Petra, Amman, Jordan*
[3]*Department of Chemical Engineering, Birla Institute of Technology and Science Pilani KK Birla Goa Campus, Sancoale, India*
[4]*Department of Mechanical Engineering, Dongguk University, Seoul, South Korea*
[5]*Université Clermont Auvergne, Clermont Auvergne INP, CNRS, Institut Pascal, Clermont-Ferrand, France*
[6]*Dept of Food Technology, Kongu Engineering College, Erode, Tamil Nadu, India*

Abstract

This chapter examines the multifaceted role of green engineering (GE) within the circular economy (CE) framework. It starts by outlining the foundational principles of sustainable design and extends to practical applications in urban development and industrial processes. Case studies demonstrate how these principles effectively minimize environmental impact while promoting economic viability. The chapter emphasizes the synergies between GE and the CE, highlighting the importance of interdisciplinary collaboration and innovative technologies in enhancing resource efficiency and sustainability. These interactions not only support environmental goals but also foster economic growth and social well-being, showcasing the comprehensive benefits of GE practices. The chapter underscores the necessity of ongoing innovation to address the escalating environmental challenges posed by modern industrial activities and urban development. It stresses

Corresponding author: gayathrichem03@gmail.com
†*Corresponding author*: vivekr@goa.bits-pilani.ac.in

Senthilkumar Rathnasamy, Vivek Rangarajan and Umile Gianfranco Spizzirri (eds.) Green Engineering and Sustainable Technology: Tools and Techniques, (1–44) © 2026 Scrivener Publishing LLC

the importance of translating sustainable practices from theoretical frameworks to real-world applications, achieving measurable environmental, economic, and social benefits. Additionally, the role of policy, education, and corporate responsibility in fostering the adoption of GE is highlighted, advocating supportive regulatory frameworks and incentives. Looking ahead, GE is poised to become increasingly influential as global resource pressures intensify. Emerging technologies such as artificial intelligence, Internet of Things, and advanced material sciences promise breakthroughs in sustainable design and production. As circular principles of the CE become more integrated into global strategies, GE will shape resilient infrastructure and systems, positioning itself as a vital component of sustainable development worldwide.

***Keywords*:** GE, CE, sustainable design, environmental sustainability, resource efficiency, urban development, industrial processes, renewable materials

1.1 Definition of Green Engineering

Green engineering (GE) is an innovative discipline that represents a significant advancement in the way that engineering is approached and practiced. It directly responds to the increasing global need for sustainable development, aiming to minimize environmental impacts while maximizing efficiency. This is achieved by integrating life cycle thinking into all engineering activities, from the initial design phase to decommissioning. The field of GE encompasses a very wide range of principles and practices applied across various sectors such as chemical manufacturing, materials processing, and product design. Advanced technologies and methodologies are leveraged to reduce waste, conserve resources, and enhance system efficiencies. This approach plays a crucial role in addressing the pressing issues of climate change, resource depletion, and ecological degradation [1, 2].

The significance of GE cannot be overstated, as it provides essential interventions for industrial practices that align with the earth's ecological boundaries. This chapter will delve into the foundational definitions and scope of GE, underlining its pivotal role in fostering sustainable practices [3, 4]. Additionally, it will explore how GE integrates within the framework of a circular economy (CE), supporting sustainable, closed-loop industrial systems that are essential for achieving sustainable development goals. Through this comprehensive exploration, the chapter aims to provide an in-depth understanding of the potential and challenges of GE, emphasizing its significance in the broader context of modern engineering practices.

The exploration of innovative practices in GE not only tackles environmental issues but also brings substantial economic advantages by enhancing efficiency and driving cost reductions across various industries [5]. Embracing these principles can lead to significant progress toward reducing energy consumption, minimizing waste, and optimizing the use of raw materials, all of which are pivotal for both sustainability and profitability. Integrating GE principles into core business strategies enables companies to bolster their competitive edge while ensuring compliance with regulations and promoting environmental stewardship. This chapter will explore the practical applications of GE, demonstrating the successful implementation of theoretical principles through real-world case studies. By exploring both achievements and ongoing challenges, the chapter aims to provide an unbiased perspective that advocates for adopting GE practices while shedding light on the complexities and trade-offs associated with transitioning to greener methodologies in traditional industries.

The evolution of GE has been marked by significant milestones, beginning with the emergence of environmental awareness in the period of 1960s to 1970s. This period saw the introduction of regulations aimed at reducing industrial pollution and promoting sustainability. Over the decades, advancements in technology, increased focus on renewable energy, and the integration of CE principles have continually shaped and expanded the field, leading to innovative practices that prioritize environmental protection and resource efficiency. Figure 1.1 shows the timeline illustrating the historical development and key milestones in the evolution of GE.

As an evolving discipline, GE encompasses a range of methodologies aimed at promoting sustainability across diverse engineering practices, significantly influencing how industries approach the environmental impacts of their processes. The concept of GE emerged in response to the increasing recognition of the necessity for sustainable development and the environmental degradation associated with traditional engineering practices. Formally defined in the early 1990s, it is rooted in the principles of reducing waste and pollution at the source and improving the efficiency of resource use throughout the lifecycle of processes and products. This historical evolution represents a substantial shift from reactive environmental measures to proactive integrated solutions in engineering design and implementation [6].

The concept of GE extends beyond the traditional focus on pollution control. It encompasses the development of systems and processes that are inherently environmentally sustainable. Different industries apply these principles in accordance with their specific environmental impacts and

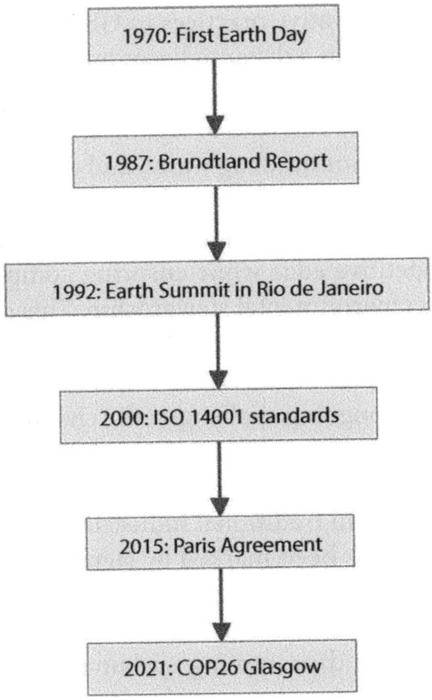

Figure 1.1 The evolution of green engineering.

operational contexts. For instance, in the chemical industry, GE entails designing processes that minimize the production of toxic substances and maximize energy efficiency. In civil and construction engineering, the emphasis often lies in the utilization of sustainable materials and the optimization of energy efficiency [5] in buildings and infrastructures [7]. This demonstrates the adaptability of GE to diverse industry needs, emphasizing that it is not a one-size-fits-all approach.

Furthermore, it is becoming increasingly apparent that companies are recognizing the significance of incorporating GE principles into their operations to not only comply with regulatory standards but also satisfy consumer preferences for sustainable practices [8]. In fact, implementing GE principles in the current industry might face, on short time interval, a type of resistance from the owners of factories as they already have a well-established production procedure, they have cheap synthetic raw materials and resources, and the infrastructure cost of their current factories and procedures is already available. However, such cost challenges might convert into opportunities, and the production cost especially the energy expenses will reach the minimum on long time intervals as long as the GE principles are applied. This strategic integration has been shown to yield benefits such as cost reductions, enhancement of brand reputation, and a competitive advantage within the market. Consequently, the incorporation of GE principles is not solely an ethical obligation but also a strategic business move that can profoundly impact a company's long-term sustainability. Building on our earlier discussion regarding the emergence of GE, it is evident that the pervasive adoption of its principles is progressively evolving into a cornerstone of companies' fundamental strategies on a global scale. This integration signifies a substantial departure from conventional engineering paradigms, signaling a profound shift toward a more sustainable and environmentally conscious approach. This shift closely aligns with the overarching objectives of the CE and sustainable development goals. Nonetheless, the persistent challenge lies in standardizing these practices across diverse geographical and industrial contexts, ensuring the universal adoption and effective implementation of GE principles.

Despite the growing adoption and clear benefits of GE, several myths persist that can hinder its broader acceptance and implementation. One common misconception is that GE practices are inherently costlier than traditional methods. This belief overlooks the long-term savings through energy efficiency, waste reduction, and improved process efficiencies. Additionally, there is a misconception that these practices compromise on performance or product quality. In fact, GE often enhances performance by optimizing resource use and incorporating innovative technologies. Another myth is that GE is only applicable to new projects or industries focused on environmental products, whereas it is relevant and can be integrated into existing systems across all sectors. These myths create unnecessary barriers to adoption, underscoring the need for continued education and demonstration of GE's practical benefits and versatility in addressing a wide array of engineering challenges. By dispelling these misconceptions, the engineering community can further embrace sustainable practices that align with both economic and environmental goals [9].

GE is a transformative, multidisciplinary approach that transcends traditional environmental engineering by integrating principles of sustainability into various sectors, including industrial design, urban planning, and waste management. In industrial design, GE drives the development of products and processes that prioritize environmental friendliness, energy efficiency, and sustainability throughout their entire lifecycle. This approach aims to minimize waste generation and reduce environmental impact while meeting the growing consumer demand for eco-friendly and sustainable products. Moreover, adopting green design principles can offer significant competitive advantages to companies. According to researchers, businesses that embrace these principles often experience improved operational efficiencies and demonstrate compliance with global environmental standards, positioning themselves as leaders in sustainable business practices [10].

In the field of urban planning, GE principles play a crucial role in creating sustainable infrastructures that prioritize energy efficiency, minimal environmental impact, and improved urban living standards. Examples of projects that embody these principles include incorporating green spaces into urban areas, implementing sustainable water management systems, and designing energy-efficient buildings. These initiatives are particularly important as urban areas are projected to expand, with estimates indicating that almost 68% of the world's population will be urban areas by 2050 [11, 12]. By incorporating GE strategies into urban development, we can significantly reduce the negative effects of urbanization on the environment and human health. Waste management is a critical area where GE has a significant impact. By advancing methods for waste reduction, reuse, and recycling, GE contributes to the efficiency of resource use and the reduction of landfill use and environmental pollution. Effective waste management practices are essential for sustainable development because they ensure that resources are cycled within economies rather than being lost as waste. Reports emphasize the role of GE in redefining waste as a resource and points out the potential for innovation in developing new materials and processes that convert waste into valuable products [13].

The interdisciplinary nature of GE sets it apart as a transformative field. It integrates concepts from various engineering disciplines with insights from environmental science, economics, and social sciences to create solutions that are not only technically viable but also economically feasible and socially acceptable. This holistic approach is vital for addressing the complex challenges of sustainable development, making GE a key player in the global pursuit of sustainability. As the pressures on our global environment intensify, the scope of GE is expected to expand further, necessitating a greater incorporation of these principles into all facets of engineering education and practice.

Accordingly, GE principles aim at providing a comprehensive framework for the successful integration of environmental considerations into engineering practices that ultimately promote the reduction of pollution at its source, the improvement of energy efficiency, and the sustainable utilization of resources. Figure 1.2 shows an example of a sustainable building that utilizes the principles of GE and is designed to be environmentally friendly and fulfill the GE principles.

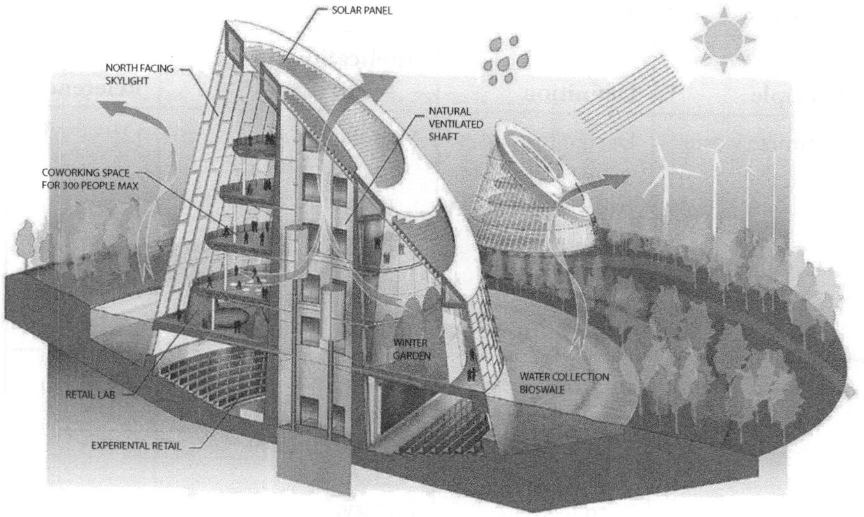

Figure 1.2 Sustainable building that utilizes the principles of GE [14].

The core principles of GE act as a comprehensive framework for the successful integration of environmental considerations into engineering practices. These principles promote the reduction of pollution at its source, the improvement of energy efficiency, and the sustainable utilization of resources. In 2003, Anastas and Zimmerman delineated 12 principles that emphasize the minimization of waste, the enhancement of energy efficiency, the reduction of non-renewable resource usage, and the maximization of renewable resource usage [4]. It is important to note that these principles are not merely theoretical concepts; rather, they represent practical guidelines that influence the entire lifecycle of a product or process, from its initial design to its eventual disposal. Recent strides in research have propelled the practical application of these principles, leading to notable outcomes. Take, for example, the significant progress in developing biodegradable materials for industrial design, a direct response to the imperative of reducing environmental impact through pioneering material science. Turner and fellow researchers have showcased how these biodegradable composites can effectively supplant traditional plastics across various applications, thus substantially mitigating the long-term environmental risks associated with waste disposal [15]. Similarly, the energy sector has witnessed remarkable advancements in enhancing the effectiveness and sustainability of power systems. Breakthroughs in solar panel technologies and the seamless integration of smart grids serve as prime illustrations of

Table 1.1 Key principles of green engineering.

Principle	Definition	Applications in sectors	References
Minimize waste	Design processes that minimize the volume and toxicity of waste throughout the life cycle	Manufacturing: design for assembly to reduce scraps Construction: use of prefabricated components	[2]
Maximize energy efficiency	Improve energy efficiency and give preference to using renewable energy sources	Automotive: development of more efficient engines Energy: implementation of smart grid technologies	[20]
Conserve natural resources	Use renewable or recyclable materials to reduce the depletion of natural resources	Textiles: use of organic or recycled materials Electronics: emphasize design for simplified disassembly and recyclability	[21]
Extend product lifespan	Design for longevity and durability, encouraging products to be used to their fullest before disposal or recycling	Consumer goods: modular designs that can be easily upgraded Technology: longer-lasting electronic devices	[22]

(Continued)

Table 1.1 Key principles of green engineering. (*Continued*)

Principle	Definition	Applications in sectors	References
Use life cycle thinking	Evaluate the environmental effects from extracting raw materials to disposal to minimize our ecological footprints	Packaging: biodegradable materials Agriculture: sustainable crop rotation practices	[2]
Innovate	Drive innovations that enhance sustainability without compromising productivity and usability	Pharmaceuticals: green synthesis methods Architecture: innovative use of sustainable building materials	[23]
Integrate design for disassembly	Design products in a manner that facilitates the easy separation and reusability of each component upon reaching the end of the product's life cycle	Furniture: easy disassembly and component replacement Vehicles: modular components for easier recycling	[24]

(*Continued*)

Table 1.1 Key principles of green engineering. (*Continued*)

Principle	**Definition**	**Applications in sectors**	**References**
Reduce resource consumption	Design processes and products that use less raw material and energy to achieve the same or better functionality	Industrial equipment: high- efficiency machinery Home appliances: energy-efficient appliances	[22]
Enhance system adaptability	Develop systems and processes that can adapt to use alternative and renewable resources without requiring major changes	Power systems: flexible fuel use capabilities Water management: systems that adapt to various water sources	[2]
Promote environmental risk management	Prioritize risk assessment and management strategies to mitigate environmental impacts	Chemical industry: safer chemical processes Mining: reduced environmental footprint techniques	[21]
Encourage environmental advocacy	Foster a corporate and community culture that values and promotes environmental awareness and actions	Corporate policies: sustainability as a core value Community programs: local green initiatives	[22]

GE principles in action, as they precisely optimize energy usage and prioritize renewable energy sources over their non-renewable counterparts. Although the preparation of solar panels requires a lot of energy and leads to an increase in the total entropy of the universe, it has significant economic outcomes in reducing the energy expenses and providing a way for replacing renewable with non-renewable energy source [16].

Furthermore, the application of GE principles in urban planning has seen substantial improvements in the sustainability of urban environments. Sustainable urban drainage systems and green buildings are practical examples where these principles are implemented to manage water resources more sustainably and increase the energy efficiency of structures. Research highlights how urban projects that integrate green roofs and walls not only enhance the aesthetic value of the cityscape but also contribute significantly to the reduction of urban heat islands, thus improving urban air quality and reducing the overall environmental footprint of cities [17, 18].

The profound and ever-expanding influence of GE on environmental sustainability cannot be overstated. As dedicated researchers continually push the boundaries of what is achievable from both a technical and economic perspective, the fundamental principles of GE are steadily finding their way into broader systems and processes. This integration is playing a pivotal role in cultivating a more sustainable industrial landscape, fostering resilience and adaptability across a wide spectrum of sectors. The ongoing research and practical application of GE principles signify a transformative shift toward a more sustainable future, vividly illustrating the potential for engineering solutions to seamlessly blend industrial progress with environmental stewardship.

Here is a conceptual design for Table 1.1, summarizing the core principles, their definitions, and how they apply across various sectors. This table aims to condense the broad scope of GE, demonstrating its relevance and implementation across diverse industries.

1.2 Definition of Circular Economy

The idea of a CE represents a significant shift in economic development, aiming to bring benefits to businesses, society, and the environment. In contrast to the conventional linear economy, which operates on a "take-make-dispose" model, a CE focuses on eliminating waste and pollution, prolonging the use of products and materials, and rejuvenating natural systems [25–27]. In other words, CE is going to connect the last word of

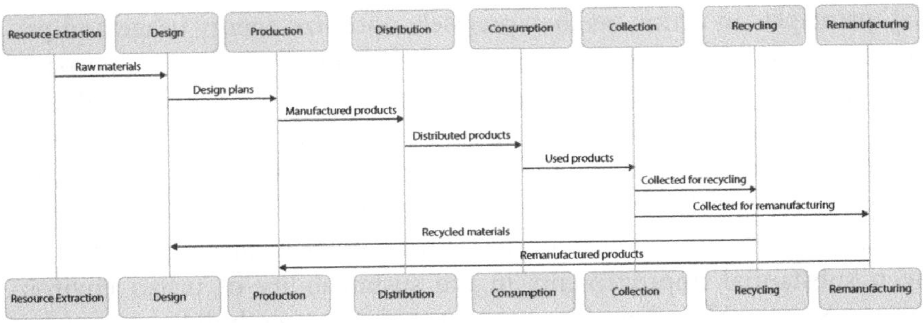

Figure 1.3 The circular economy model.

"take-make-dispose" model to the first and convert the relation into circular instead of linear. This model not only aims to achieve economic stability but also tackles important environmental issues such as resource depletion, waste, and pollution. The origins of the CE concept can be traced back to several key theories and schools of thought including the blue economy, biomimicry, and cradle-to-cradle design. These introductory ideas began to coalesce in the late 1970s and early 1980s, as researchers like Pearce and Turner (1989) started advocating for a closed-loop economy, influenced by earlier notions of sustainable development and systems thinking. The formal term "CE" gained prominence in the 21st century as it was adopted by several international organizations and businesses, driven by the increasing need to address global environmental challenges.

A diagrammatic representation of the CE model visually illustrates how resources flow in a closed loop, emphasizing the key components of reduce, reuse, recycle, and regenerate. It showcases how materials are designed to cycle through multiple stages, minimizing waste and maximizing resource efficiency, thus promoting sustainability and environmental stewardship. Figure 1.3 is a diagrammatic representation of the CE model, showing its key components and flow of materials.

1.2.1 Linear vs. Circular Economies: Key Principles and Implications

The distinctions between linear and circular economies have become pivotal in the realm of sustainable development, underscoring the comprehensive approach of GE to integrating sustainability across various sectors. Traditionally, the linear economy adheres to a "take-make- dispose" model, a process characterized by a straightforward sequence where raw materials are transformed into products that are used until they are eventually discarded as waste. This model has been criticized for its unsustainable consumption of finite resources and substantial waste generation, leading to environmental degradation [26]. Conversely, the CE proposes a regenerative approach designed to retain as much value as possible from products, parts, and materials. This model emphasizes minimizing waste through the continual use of resources, incorporating recycling, reuse, repair, and remanufacturing processes, thereby extending the lifecycle of products [34]. By designing out waste and reducing the environmental impacts of production, the CE helps mitigate resource scarcity and environmental pollution. Furthermore, it fosters innovation and opens up new business opportunities through alternative economic models such as product-as-a-service [36].

The shift to a CE is seen not only as an environmental imperative but also as a strategic economic and social development mechanism. It promises a more sustainable and resilient future by encouraging the development of systems and processes that are inherently environmentally sustainable. Different industries apply these principles in accordance with their specific environmental impacts and operational contexts. For instance, in the chemical industry, GE entails the design of processes that minimize the production of toxic substances and maximize energy efficiency [35]. In civil and construction engineering, the emphasis often lies in the utilization of sustainable materials and the optimization of energy efficiency in buildings and infrastructures [26]. This adaptability of GE to diverse industry needs emphasizes that it is not a one-size-fits-all approach. By integrating GE principles, industries can enhance their ability to reduce the environmental impact of their operations while creating economic value through resource efficiency. This synergy allows for a cohesive approach to sustainability, where each discipline reinforces the goals of the other, creating a comprehensive model for environmental and economic sustainability. Such practices are increasingly documented in the literature, illustrating how industries from manufacturing to construction are reaping the benefits of this integrative approach [33]. The shift from a linear to a CE, facilitated by GE, is essential for preserving the environment and ensuring economic viability. This transition represents not only a response to environmental challenges but also a strategic opportunity for innovation and sustainable development, promising a more resilient and sustainable future.

As presented below, Table 1.2 outlines the key characteristics of both economic models and highlights the benefits and implications of adopting a circular approach [33].

Table 1.2 Comparison of linear and circular economies [5, 23, 38, 39].

Characteristic	Linear economy	CE
Resource use	Uses resources in a "take-make- dispose" pattern. Resources are used once	Emphasizes the reuse, repair, refurbishment, and recycling of resources
Environmental impact	High waste generation and pollution due to single-use focus and disposal	Significantly reduced waste and pollution through continuous use of resources
Economic model	Cost-driven; focused on short-term gains from product sales	Value-driven; focuses on long-term resilience and sustainability of resources
Product design	Products designed for obsolescence and disposability	Products designed for durability, reparability, and recyclability
Supply Chain	Linear supply chain: raw materials → production → consumer → landfill	Circular supply chain: resource recovery and continuous flow of materials
Innovation Focus	Driven by market demands, often at the expense of environmental considerations	Driven by sustainability, aiming for technological innovations that support reuse
Business Model	Traditional business models sell products as the endpoint of the value chain	Innovative business models, such as product-as-a-service, that extends product life
Economic Implications	Potential resource shortages due to overconsumption and dependency on raw materials	Economic stability through resource efficiency and reduced dependency on raw materials
Social Implications	Lesser focus on social impacts or community involvement in production cycles	Promotes community engagement and local job creation through recycling and maintenance activities

The shift from a linear to a CE involves not just changing how resources are used but transforming the entire mindset around production, consumption, and waste. The CE model facilitates sustainable development by integrating environmental health, economic viability, and societal well-being. Adopting this model can lead to reduced environmental impact, enhanced resource security, and improved economic stability through more efficient practices. As highlighted in the table, the CE also fosters innovation and creates opportunities for new business models that can adapt and thrive in a resource-constrained world. This table aims to provide a clear, structured comparison that underscores the urgency and benefits of transitioning toward a CE, particularly in the context of global sustainability challenges [21, 23, 38, 40–43].

1.2.2 Historical Development and Global Adoption of CE

The adoption of CE principles has evolved from theoretical frameworks to implementation at the global, national, and local levels. The European Union has been a pioneer in this regard, integrating CE strategies into its environmental and economic policies, notably through its CE Action Plan, which includes measures to encourage recycling, sustainable consumption, and waste management practices [11–25]. Countries like China have also implemented policies to promote CE practices, integrating them into national industrial policy to enhance resource efficiency and reduce environmental pollution [28].

Research and development have played a critical role in advancing the CE. Recent innovations include the development of biodegradable materials, advanced recycling technologies, and business models that encourage the reuse and refurbishment of products. Researchers continue to explore the implications of circular strategies on various sectors and evaluate their potential to mitigate climate change, promote biodiversity, and conserve resources. Such CE idea agrees with the law of conservation of mass that says matter can be neither created nor destroyed and the role of GE and CE is then to understand the process of matter conversion and trying to control it to make the waste reproducible. The global adoption of the **CE** is not only a testament to its viability but also an indicator of the paradigm shift in how societies value and utilize resources.

1.3 Key Principles of the Circular Economy

1.3.1 Design Out Waste and Pollution

The principle of designing out waste and pollution is fundamental to the CE, challenging the traditional take-make-waste model by embedding sustainability into the very fabric of product design and production processes. This approach seeks to ensure that products are designed from the outset to minimize waste and environmental impact through strategies such as using non-toxic, renewable, or recyclable materials and designing for product longevity and reparability [17]. Researchers have emphasized the importance of integrating environmental considerations in the early stages of product development to reduce resource consumption and prevent waste streams from forming. In practice, advances in material science have led to the creation of composites and biopolymers that are easier to recycle and degrade, thus minimizing pollution and waste associated with product disposal [29]. Moreover, the adoption of modular design principles allows for easier disassembly and refurbishment, thereby extending the lifecycle of products. These innovations are not only reducing environmental burdens but also reshaping consumer behaviors and business models toward more sustainable practices. The impact of this principle on the environment is profound. By reducing the waste generated during production and at the end of a product's life, industries can significantly decrease their ecological footprint. This not only conserves resources but also cuts down on pollution, contributing to cleaner air, water, and soil.

1.3.2 Keep Products and Materials in Use

The principle of keeping products and materials in use revolves around maximizing the utility of resources by maintaining, repairing, refurbishing, and recycling materials to extend their lifecycle. This approach underpins the shift from a disposable culture to one that values longevity and sustainability [26]. It encourages the design of products with longer lifespans and supports systems that facilitate repair and reuse, reducing the demand for raw materials and decreasing waste.

Recent research has focused on developing and implementing business models such as product-as-a-service, where goods are leased rather than sold—this not only encourages manufacturers to create durable and maintainable products but also ensures that the products are used more efficiently [30]. Additionally, technological advancements in recycling technologies have improved the ability to reclaim and reuse materials, even from complex products like electronics and batteries, thereby keeping them within the economic cycle for as long as possible. The environmental benefits of this principle are clear: reducing material throughput in the economy decreases pressure on natural resources and lowers greenhouse gas emissions associated with production and disposal. This practice not only supports sustainability but also enhances energy efficiency and fosters a more resilient economic system.

1.3.3 Regenerate Natural Systems

Regenerating natural systems is a transformative principle of the CE that extends beyond minimizing negative impacts to actively improving the environment. This principle emphasizes restoring, renewing, and revitalizing sources of natural capital, aiming to leave the environment in a better state than it was found. It involves adopting practices that enhance ecosystem services and biodiversity, such as using regenerative agricultural techniques or restoring natural habitats to improve soil health and water quality [31, 32]. Current research in this area includes the development of green infrastructure and urban planning initiatives that integrate nature-based solutions to address urban environmental and social challenges. Projects like creating urban green spaces, wetlands, and permeable surfaces not only help manage stormwater and reduce urban heat island effects but also provide habitats for wildlife and recreational spaces for communities [33]. The potential of this principle to impact the environment positively is immense. By actively enhancing the natural systems upon which economic activities rely, the CE supports a sustainable coexistence between human enterprises and the Earth's ecological systems, ensuring long-term sustainability and resilience.

1.4 Circular Economy as Critical Factor of Green Engineering

GE plays a pivotal role in driving the principles of the CE by designing systems and processes that emphasize sustainability, efficiency, and the minimization of waste. By integrating GE principles, industries can significantly enhance their ability to not only reduce the environmental impact of their operations but also create economic value through resource efficiency. GE's approach to optimizing system designs for reduced material usage and extended product lifecycles directly supports the CE's aim to maintain the value of products, materials, and resources for as long as possible [34, 35]. In practice, GE facilitates the CE through innovations such as designing for disassembly, where products are designed thoughtfully to ensure that they can be easily taken apart. This allows for more effective recycling and remanufacturing, which are core aspects of the CE. For instance, advancements in modular construction in the automotive and electronics industries have shown how design innovations can lead to significant improvements in product lifecycle management and material recovery [36]. Moreover, the integration of GE practices into industrial processes has led to the development of cleaner production technologies that not only reduce waste but also decrease the consumption of raw materials and energy. This is particularly evident in the chemical industry, where GE has revolutionized the development of processes that maximize the use of renewable resources and minimize toxic byproducts, thereby supporting circular economic models [27]. By employing GE, industries are not only advancing their sustainability goals but are also enhancing their economic performance through increased resource efficiency and reduced waste. This synergy between environmental and economic benefits underscores the vital role of GE in realizing the vision of a CE.

Figure 1.4 shows a flowchart depicting how the principles of GE support and enhance the CE. This visually emphasizes how GE practices play a vital role in achieving sustainable production and consumption patterns essential for a CE.

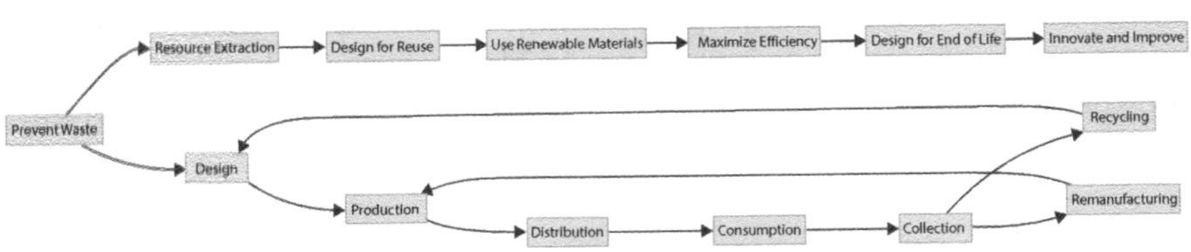

Figure 1.4 Green engineering principles and circular economy.

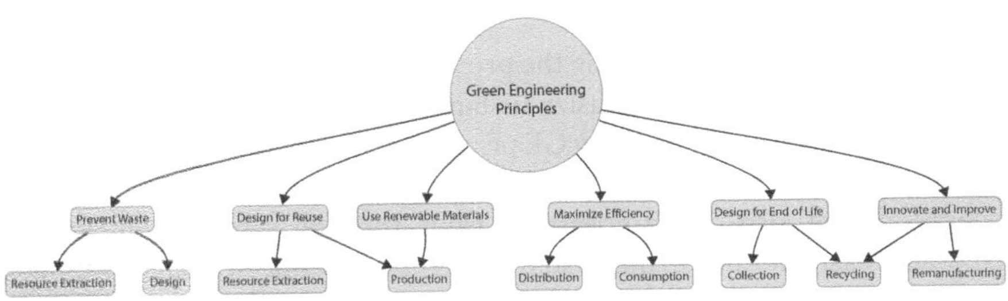

Figure 1.5 Industrial aspects of circular economy.

Figure 1.5 depicts the processes involved in a specific industrial case study that applies CE principles. This approach is general and can be utilized as a guide for any case study.

1.5 Circular Economy and Green Engineering Case Studies

Several case studies highlight the successful implementation of CE principles through GE. One notable example is the initiative by Philips with their "Pay per Lux" model, where they sell lighting services rather than light bulbs. This business model encourages Philips to design products that are durable, energy-efficient, and easier to maintain and recycle. Another example is Interface, Inc., a global leader in modular carpet design, which has adopted a closed-loop process where used carpets are recycled into new products. This not only reduces the demand for virgin raw materials but also decreases waste and environmental impact [26]. In the construction sector, the Park 20|20 development in The Netherlands exemplifies the application of CE and GE principles. The development uses cradle-to-cradle design principles to ensure that all materials used can be reused at their highest value, incorporating renewable energy sources and sustainable water management practices to create a fully sustainable business park [37]. These case studies demonstrate how GE is instrumental in implementing CE principles, leading to sustainable, efficient, and economically viable solutions across various industries.

Several case studies from diverse sectors vividly illustrate the successful implementation of GE principles within the framework of the CE, underscoring their broad applicability and effectiveness in promoting sustainability. In the electronics sector, Fairphone in The Netherlands epitomizes the application of modular design, product longevity, and recycling. Fairphone's modular smartphones are designed for easy repair and upgrade, utilizing recycled materials and offering recycling programs for old devices. This approach not only reduces e-waste but also promotes a more sustainable model for consumer electronics, demonstrating that tech innovation can be aligned with environmental responsibility. In the furniture industry, IKEA's global operations exemplify the integration of CE principles through initiatives such as furniture leasing programs and the use of sustainable materials like bamboo and recycled wood. By promoting the reuse and recycling of furniture, IKEA significantly reduces waste and sets a benchmark for sustainability in large-scale manufacturing.

Similarly, Patagonia in the USA showcases the effectiveness of repair, reuse, and recycle principles in the textiles sector through its Worn Wear program. By encouraging customers to repair, share, and recycle their clothing, Patagonia extends the life of apparel and diminishes the demand for new materials, highlighting a sustainable approach to fashion. The construction sector also provides compelling examples, particularly with the Park 20|20 development in The Netherlands, which employs cradle-to-cradle design and renewable energy use. This project demonstrates how sustainable urban planning and construction can lead to buildings and communities that are not only environmentally friendly but also economically viable and socially beneficial. In the automotive industry, Renault in France has pioneered a remanufacturing program where car parts like engines and transmissions are reconditioned and sold at competitive prices. This practice reduces waste and energy consumption, illustrating how remanufacturing can contribute to a more sustainable automotive industry.

Urban development projects, such as Singapore's NEWater initiative, highlight the potential for sustainable urban planning to address critical resource challenges. NEWater recycles treated wastewater into ultra-clean, high-grade reclaimed water, meeting up to 40% of Singapore's water needs. This project significantly advances the city-state's water sustainability strategy, showcasing how innovative recycling technologies can ensure resource security. These case studies collectively demonstrate the practical implementation of GE and CE principles across various industries and regions. They prove that sustainability is achievable and beneficial, offering models that other businesses and sectors can replicate. By reducing environmental impact, enhancing resource efficiency, and promoting sustainable practices, these examples underscore the comprehensive benefits of adopting GE practices, paving the way for a more sustainable and resilient global economy. Table 1.3 shows case studies on CE implementations and provides examples from various sectors demonstrating the successful application of CE principles. Each case illustrates how different industries and regions have adopted these practices, along with the outcomes of such implementations. Furthermore, each case study demonstrates the practical implementation of CE principles tailored to specific industry needs and contexts. The outcomes highlight the benefits of adopting such practices, including reduced environmental impact, enhanced sustainability, and, in some cases, economic benefits through resource efficiency. These examples serve as models for other businesses and sectors considering the transition to a CE framework. They show a broad spectrum of applications, underscoring the versatility and effectiveness of CE principles across different geographical and industrial landscapes.

Table 1.3 Case studies on CE implementations, illustrating the successful application of CE principles.

Sector	Company/ project	Location	Key principles applied	Outcomes	Ref.
Electronics	Fairphone	Netherlands	Modular design, longevity, and recycling	Fairphone designs modular smartphones that are easy to repair and upgrade They use recycled materials and offer a recycling program for old devices, reducing e-waste and promoting sustainable consumer electronics.	[19, 57, 58]
Furniture	IKEA	Global	Product as a service, sustainable materials, and recycling	IKEA has initiated a furniture leasing program that supports the reuse and recycling of products. They also use sustainable materials like bamboo and recycled wood. This approach has significantly reduced waste.	[57, 59]

(*Continued*)

Table 1.3 Case studies on CE implementations, illustrating the successful application of CE principles. (*Continued*)

Sector	Company/ project	Location	Key principles applied	Outcomes	Ref.
Textiles	Patagonia	USA	Repair, reuse, and recycle	Patagonia's Worn Wear program encourages customers to repair, share, and recycle their clothing. The initiative supports sustainability by extending the life of apparel and reducing the demand for new materials.	[19, 60]
Construction	Park 20	20	Netherlands	Cradle-to-cradle design, renewable energy use.	[61]
Urban development	Singapore	Singapore	Water recycling and sustainable urban planning	Singapore's NEWater project recycles treated wastewater into ultra-clean, high-grade reclaimed water, meeting up to 40% of the nation's water needs and significantly contributing to its water sustainability strategy.	[62, 63]

1.6 Prospects and Challenges in Synergy between Circular Economy and Green Engineering

1.6.1 Prospects of the Synergy

The intersection of GE and the CE reveals significant synergies and opportunities for collaboration, enhancing sustainability and efficiency across various industries. GE principles, such as the reduction of material use and the enhancement of energy efficiency, align closely with the goals of the CE to maintain the value of resources and minimize waste. This synergy allows for a cohesive approach to sustainability, where each discipline reinforces the goals of the other, creating a comprehensive model for environmental and economic sustainability [64].

For instance, GE's emphasis on eco-designing products with their end-of-life in mind directly supports the CE's focus on resource recovery and the continuation of resource cycles. This collaboration has led to innovations such as the development of new materials that are easier to recycle and designs that facilitate easier disassembly at the end of a product's life. Researchers in materials science and industrial design are increasingly focusing on how materials can be reused and recycled without losing their value, contributing to the development of a circular materials economy [36]. Moreover, the integration of GE practices into circular economic models has resulted in enhanced operational efficiencies. For example, the use of GE techniques in optimizing water and energy systems within manufacturing processes not only reduces the environmental impact but also lowers operational costs, demonstrating joint economic and environmental benefits. Such practices are increasingly documented in the literature, illustrating how industries from manufacturing to construction are reaping the benefits of this integrative approach [22, 45]. However, this synergy is not without challenges. Ongoing research in both fields is focused on overcoming these hurdles. Studies are exploring how policy frameworks can be designed to support the integration of GE and CE principles, ensuring that the regulatory environment encourages rather than impedes sustainable innovation. Additionally, as industries adopt these integrated practices, they provide valuable case studies and data that further refine and enhance the models and practices in use, driving continuous improvement in sustainability and efficiency on a global scale [20, 45].

Table 1.4 Challenges in integrating green engineering with circular economy.

Challenge	Description	Suggested solutions	References
Technological limitations	The lack of advanced technology can hinder the efficient processing and recycling of materials.	Invest in research and development to innovate new materials and technologies that facilitate recycling and remanufacturing	[47–49]
Economic viability	High upfront costs associated with transitioning to circular models can be prohibitive.	Develop financial incentives and support mechanisms to offset initial costs and highlight long-term economic benefits	[24, 49–51]
Regulatory and policy barriers	Inadequate regulatory frameworks can impede the adoption of circular practices.	Advocate for policy reforms that support CE initiatives and provide clear guidelines for businesses	[24, 51]
Cultural and behavioral resistance	Stakeholder reluctance to change established practices can slow or block implementation.	Implement training programs and awareness campaigns to educate stakeholders on the benefits of CE practices	[52, 53]
Supply chain complexity	Managing the logistics of a circular supply chain can be challenging due to its complexity.	Utilize digital tools like AI and blockchain for better supply chain visibility and management	[54, 55]
Technical skills gap	There is often a lack of skilled workforce knowledgeable in CE and GE.	Foster educational programs and partnerships with academic institutions to train a new generation of skilled professionals	[56]

1.6.2 Challenges in Synergy

Despite the potential benefits, integrating GE into the CE presents several challenges. One of the primary hurdles is the economic viability of transitioning from linear to circular models, which often requires substantial upfront investments in new technologies and redesigning products and processes. Moreover, the lack of standardized regulations and incentives for CE practices can slow adoption rates among industries [44]. Such standard regulations have to be widely approved and applied in order to keep our earth sustainable and for the benefit of all humanity.

To overcome these challenges, researchers and policymakers are exploring various solutions, including the development of economic incentives and regulatory frameworks that encourage companies to adopt CE practices. For instance, extended producer responsibility regulations have proven effective in some regions by making producers responsible for the disposal of products at the end of their life, thereby encouraging them to design for longevity and recyclability [34]. Additionally, technological innovations and interdisciplinary collaborations are essential for addressing the technical challenges of implementing CE models [35, 45, 46]. Ongoing research into new materials, product design strategies, and industrial symbiosis networks can provide the necessary tools and frameworks for the successful integration of GE and CE principles.

Table 1.4 outlines the various challenges that industries may face when trying to integrate GE with CE principles. It provides detailed descriptions of these obstacles and proposes specific solutions to address them. Additionally, the table includes references to support further reading and verification, which are drawn from recent literature in this field. The main goal of this table is to offer a comprehensive and structured overview of the challenges associated with the integration of GE and CE principles. Each entry is supported by suggested practical solutions, making it a valuable resource for industry leaders, policymakers, and educators. By providing a foundation for further investigation and discussion, this table can help stakeholders understand and navigate the complexities of this integration more effectively.

1.7 Emerging Scope of Green Engineering and Circular Economy

The convergence of GE and the CE is set to be profoundly influenced by emerging technologies that promise to redefine sustainability and resource efficiency boundaries. Technologies such as artificial intelligence (AI), the Internet of Things (IoT), and advanced robotics are transforming traditional production and consumption patterns, offering unprecedented opportunities to optimize resource use and enhance product lifecycles [65]. AI and IoT, for instance, are crucial for developing smart systems for real-time resource management, which can significantly reduce waste and improve recycling and recovery processes. These technologies enable precise tracking and management of resources, leading to more efficient use and less environmental impact.

1.7.1 Scope of Green Engineering in Biotechnology

Biotechnology is another rapidly advancing field playing a crucial role in GE and the CE. Innovations in enzyme design and microbial processes are enabling more efficient bio-recycling of materials previously considered non-recyclable. These technologies pave the way for bio-based production methods that use renewable resources to create biodegradable products, thus supporting the CE's emphasis on minimizing waste [33, 36]. For example, bio-based plastics and other biodegradable materials are increasingly being developed to replace conventional plastics, which are challenging to recycle and often end up as environmental pollutants. Moreover, advancements in material science, particularly in developing new composites and biodegradable materials, continue to drive the principles of both GE and the CE forward. These materials are designed to have minimal environmental impact at the end of their life, either by being completely recyclable or by naturally degrading in an environmentally benign manner. The integration of these materials into product design and architecture underscores a shift toward sustainability that is both innovative and essential for future developments. Products made from these advanced materials can significantly reduce the ecological footprint by ensuring that materials are kept in use for as long as possible and are easy to recycle or biodegrade when their use phase ends.

1.7.2 Scope of Green Engineering in Industries

Looking ahead, the fields of GE and the CE are expected to evolve further, driven by global sustainability targets and increasing environmental regulations. One prediction is that there will be a greater emphasis on systems integration, where CE principles are embedded at every level of production, from individual components to entire supply chains [34]. This systemic approach will likely be supported by digital platforms that facilitate the sharing of information and resources across industries and sectors, enhancing the efficiency and scalability of circular practices. For instance, blockchain technology could be used to create transparent supply chains, ensuring that all materials are sourced, used, and recycled responsibly. Another likely trend is the increased adoption of CE models in developing economies, where rapid industrialization and urbanization present unique challenges and opportunities for highly sustainable development. As these regions invest in new infrastructure and industrial capacities, integrating CE principles from the outset could lead to more sustainable growth models that are less reliant on resource extraction and waste generation [17, 66]. This proactive approach can help these economies leapfrog to more sustainable practices, avoiding the pitfalls of the linear economy.

Finally, as public awareness of environmental issues grows, consumer demand for sustainable products and practices is expected to increase, encouraging companies to adopt GE and CE principles not only as a regulatory compliance strategy but as a central component of their business model. This shift is likely to spur further innovations and investments in sustainable technologies, reinforcing the global movement toward a more resilient and environmentally responsible economy. Consumers are becoming more conscious of their environmental impact and are seeking products that align with their values, pushing companies to innovate and prioritize sustainability in their operations. The future of GE and the CE is bright, with emerging technologies and increasing global awareness driving significant advancements. As these fields continue to evolve, they promise to offer sustainable solutions that enhance resource efficiency, reduce environmental impact, and support economic viability. The integration of these principles into mainstream practices will be critical in addressing global sustainability challenges and ensuring a resilient future for all.

1.7.3 Scope of Green Engineering in AI and IoT

Looking forward, the field of GE is set to become even more influential as global pressures on resources intensify. The integration of emerging technologies such as AI, the IoT, and advanced material sciences is likely to lead to breakthroughs that could redefine what is possible within sustainable design and production. These technologies promise to enhance efficiency, optimize resource use, and create smarter, more sustainable systems. For instance, AI can significantly improve energy management in buildings, whereas IoT can enable real-time monitoring and management of resource use across various sectors. Advanced material sciences continue to push the boundaries with innovations such as biodegradable materials and high-efficiency renewable energy systems, which are essential for reducing the environmental impact of industrial activities. Additionally, the increasing public and governmental focus on climate change and sustainability is expected to drive further investments and innovations in GE. Governments worldwide are setting ambitious targets to reduce carbon emissions and promote sustainable practices, creating a favorable environment for the development and implementation of GE solutions. This heightened focus on sustainability is also influencing consumer behavior, with more individuals and businesses seeking eco-friendly products and practices, thereby driving market demand for sustainable innovations. The future of GE also hinges on its ability to adapt to and integrate with other sustainability and technological trends. As CE principles become increasingly embedded in global economic and development strategies, GE will play a critical role in shaping resilient infrastructures and systems that not only preserve but enhance the natural environment. This integration involves designing products and processes that maximize resource efficiency, minimize waste, and support the continual use of materials through recycling, reuse, and remanufacturing. Moreover, the continual refinement of these practices, driven by research and real-world applications, promises a dynamic and impactful future for GE. Ongoing research is essential for developing new technologies and methods that can address the evolving challenges of sustainability. Real-world applications and case studies, such as those of Freiburg, IKEA, and Dell Technologies, provide valuable insights and models that can be replicated and scaled across different sectors and regions. In conclusion, the future of GE is bright, with significant potential to contribute to global sustainability efforts. Its role in integrating emerging technologies, adapting to new sustainability trends, and driving continuous innovation makes it an essential component of sustainable development worldwide. As global challenges such as resource scarcity and climate change intensify, the advancements in GE will be crucial for creating a more sustainable and resilient world, benefiting both the environment and human society.

1.8 Explicit Socio-Economic Contributions of Green Engineering

1.8.1 Industrial Application

One exemplary instance of a company that has successfully integrated GE within a CE framework is the global furniture giant, IKEA. IKEA's commitment to sustainable practices is well-documented and reflects a comprehensive adoption of CE principles, particularly in its use of sustainable materials and initiatives aimed at extending the lifecycle of its products. The company's approach includes designing products for longevity, ease of disassembly, and recycling. Additionally, IKEA has initiated take-back programs that encourage consumers to return used furniture in exchange for vouchers, effectively closing the product lifecycle loop [67]. IKEA has also made significant strides in incorporating renewable materials into its products. For example, the company has been working to replace conventional wood and plastic with recycled and renewable materials. This shift is part of IKEA's broader strategy to become "climate positive" by 2030, an ambition that also includes substantial reductions in its carbon footprint through improvements in energy efficiency and the adoption of renewable energy sources across its supply chain [68, 69]. By using such materials, IKEA not only supports sustainability but also sets a benchmark in the industry for how large-scale manufacturing can minimize environmental impacts. Moreover, IKEA's research collaborations have played a crucial role in advancing these sustainable practices. The company has partnered with numerous research institutions and technology developers to innovate new materials and recycling technologies. These collaborations have led to the development of better composites for furniture that are both recyclable and of a quality that meets consumer expectations. By investing in these technologies, IKEA is pushing the envelope on what can be achieved in product sustainability within the retail sector [70].

These initiatives by IKEA are an excellent example of how principles of GE and CE can be harmoniously integrated within industrial applications. They not only highlight the company's commitment to sustainable practices but also reflect a shift in industry standards toward more environmentally responsible production techniques. Such case studies are pivotal as they provide scalable models for other industries to follow, potentially leading to significant environmental benefits on a global scale. By demonstrating that sustainability and profitability can go hand in hand, IKEA serves as a powerful example of how businesses can lead the way in the transition to a more sustainable and CE.

1.8.2 Urban Development

Freiburg, Germany, stands as a paradigmatic example of how cities can embed GE principles into urban planning to foster sustainable living environments. Recognized as the "Green City," Freiburg has been at the forefront of ecological urban development since the 1970s. The city's comprehensive planning strategy includes extensive public transportation networks, renewable energy initiatives, and sustainable building codes, which collectively reduce the urban carbon footprint significantly [71–77]. One of the most notable aspects of Freiburg's sustainable urban development is the Vauban district, a flagship residential project. In Vauban, all homes are built to low-energy housing standards, significantly reducing energy consumption. Additionally, car ownership is highly discouraged, and superfluous road space has been repurposed for communal use and green spaces. This not only reduces traffic congestion and pollution but also promotes a sense of community and shared responsibility for the environment. Freiburg's approach to urban sustainability is deeply integrated with its social, economic, and environmental policies, which are designed to reinforce each other. The city has invested heavily in solar technology, becoming a hub for solar research and industry. This integration of renewable energy technology with urban planning showcases an innovative model of how cities can lead in the transition to a sustainable future. By prioritizing pedestrian pathways and cycling lanes, Freiburg not only improves the urban air quality but also enhances the quality of life for its residents. These measures collectively set a high standard for sustainable urban development worldwide [73, 74]. The city's efforts are a testament to the power of holistic planning and community involvement in urban development. Freiburg's success has inspired other cities globally, proving that sustainable urban development is not only environmentally beneficial but also viable and economically supportive. The holistic nature of Freiburg's policies ensures that sustainability is not treated as an isolated goal but as an integral part of the city's fabric, affecting every aspect of urban life from housing and transportation to energy use and waste management. Freiburg exemplifies how strategic urban planning, when combined with GE principles, can create a sustainable and livable city. Its innovative approaches and successful implementation of sustainable practices have set a benchmark for cities around the world. Freiburg's model demonstrates that with commitment and community engagement, urban areas can achieve significant environmental, social, and economic benefits, paving the way for a new generation of green cities. Figure 1.6 shows a visual representation (map or aerial view) of an urban area that has integrated GE principles into its development planning.

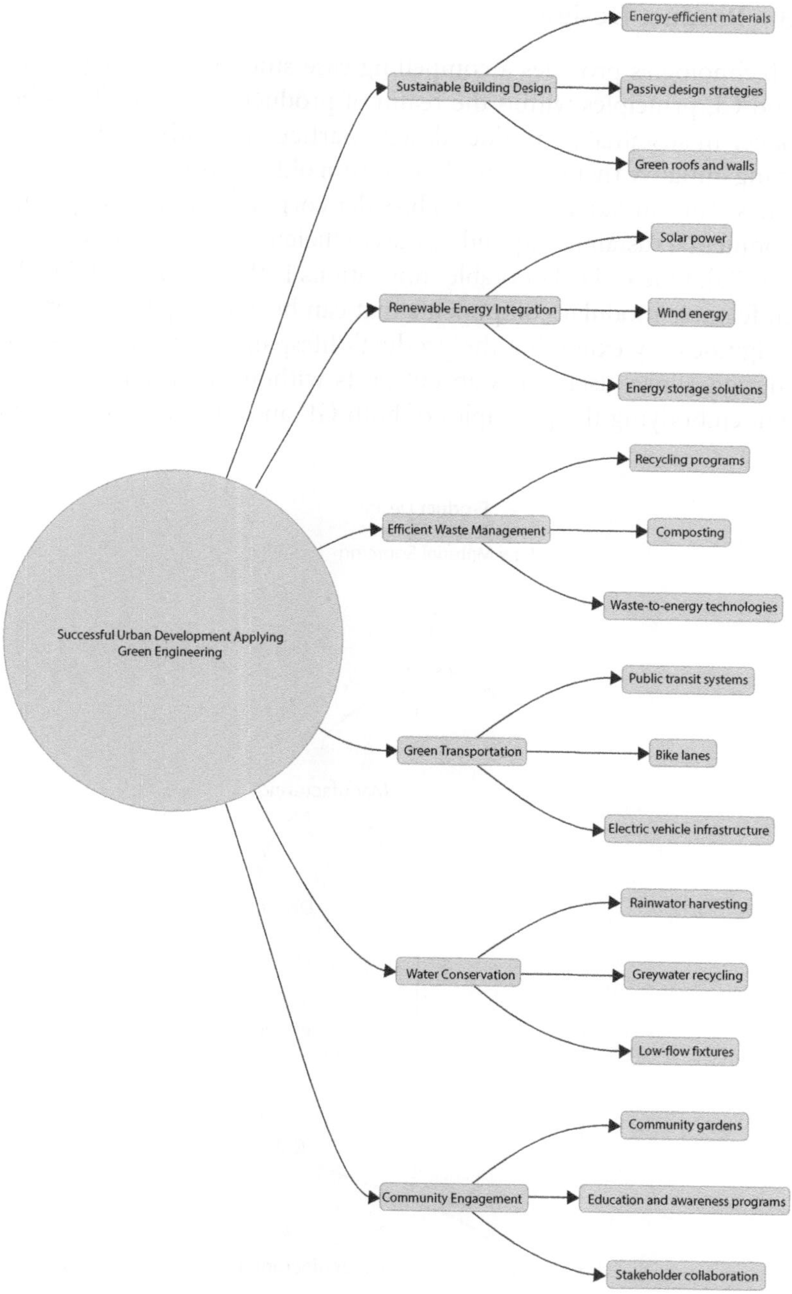

Figure 1.6 Urban development applying green engineering.

1.8.3 Product Design

Dell Technologies provides a compelling case study in the integration of GE and CE principles within the realm of product design. Dell has been a pioneer in sustainable product design, particularly with its closed-loop recycling initiative that recovers plastics from old electronics for use in new products. This initiative is part of a broader corporate strategy to promote environmental sustainability and resource efficiency across its product lines [67, 71, 75]. One of Dell's notable innovations is the OptiPlex 7070 Ultra, which features modular components that can be easily upgraded or recycled, significantly extending the product's lifespan and reducing e-waste. The design allows users to swap out parts without discarding the entire system, embodying the principles of both GE and the CE by minimizing

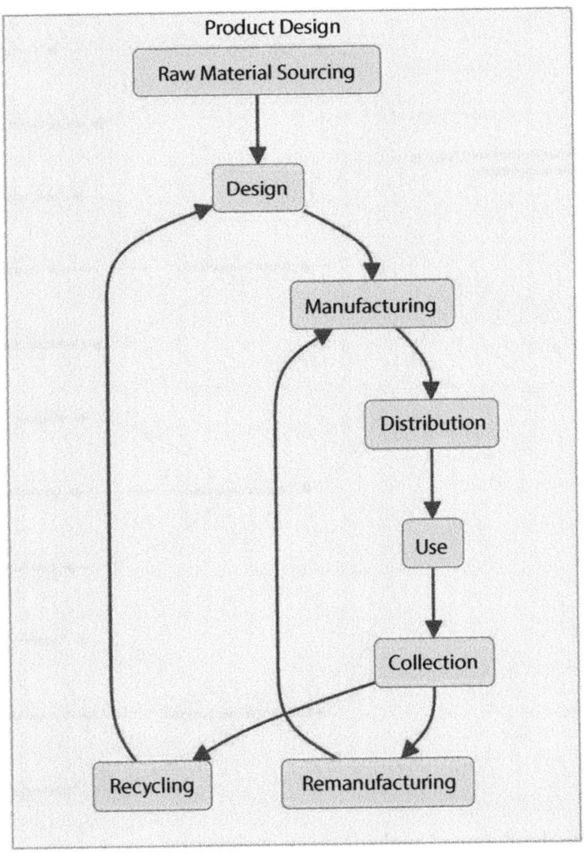

Figure 1.7 Innovative product design in circular economy.

waste and maximizing the reuse of materials [75]. This modularity not only reduces the environmental impact associated with the production and disposal of electronics but also offers economic benefits to consumers by allowing for cost-effective upgrades. Dell's commitment to sustainable product design is also evident in its packaging choices. The company utilizes innovative materials such as bamboo, mushrooms, and ocean plastics, all of which are sustainable and reduce environmental impact compared to conventional packaging materials. These packaging solutions demonstrate Dell's holistic approach to sustainability, addressing environmental impacts at every product lifecycle stage. The company's proactive approach in addressing the environmental impacts of its products from design to disposal positions Dell as a leader in sustainable product design. By incorporating GE and CE principles, Dell is not only setting new industry standards but also influencing consumer expectations toward more sustainable computing solutions. Dell's initiatives in sustainable design serve as a model for other companies, illustrating that it is possible to achieve both environmental sustainability and economic viability in the tech industry. Figure 1.7 shows a design blueprint of a product that exemplifies the integration of GE and CE principles.

1.8.4 Sustainability in Resource Management

Sustainable resources significantly influence GE by providing environmentally friendly alternatives that reduce ecological impact. Renewable energy sources like solar and wind power are pivotal in this transition, as they offer clean energy solutions that decrease dependency on fossil fuels and lower greenhouse gas emissions. For example, incorporating solar panels into urban infrastructure not only reduces the carbon footprint but also cuts energy costs, exemplifying how renewable energy enhances urban sustainability. Advances in material sciences have also led to the development of biodegradable materials derived from plant-based sources. These materials serve as eco-friendly substitutes for conventional plastics and non-renewable materials, aligning with GE principles and supporting the CE. Microbial processes and waste-to-wealth initiatives further illustrate the innovative pathways that GE can take to promote sustainability [76]. Microbial technologies, such as enzyme production for bio-recycling, enable the breakdown of complex waste materials into reusable resources. This process is highlighted in studies where microbial treatments convert agricultural and industrial waste into valuable bioproducts like biofuels and biofertilizers. These bioproducts can then be reintegrated into production cycles, transforming waste into economically valuable resources and preventing landfill accumulation. Such initiatives reflect a shift toward sustainable industrial practices, where waste is repurposed rather than discarded, demonstrating the transformative potential of GE [77]. By integrating renewable energy, biodegradable materials, and waste valorization, GE offers a comprehensive approach to environmental stewardship, fostering a sustainable future through innovative and responsible engineering practices.

Additionally, the field of GE is bolstered by the continuous development and integration of microbial and biotechnological innovations. Microorganisms, through processes like anaerobic digestion and fermentation, are being utilized to convert organic waste into bioenergy and other valuable byproducts. For instance, anaerobic digesters can transform agricultural waste and food scraps into biogas, a renewable energy source that can power homes and industries while reducing methane emissions from landfills. This not only addresses waste management issues but also provides a sustainable energy alternative. The successful implementation of such technologies is evidenced by numerous case studies, such as those in European countries where anaerobic digestion facilities have significantly contributed to national renewable energy outputs, showcasing the practical benefits and scalability of microbial solutions in GE. Moreover, waste-to-wealth initiatives exemplify the resource efficiency and economic potential inherent in GE. By repurposing industrial byproducts and consumer waste into new materials or products, these initiatives reduce raw material consumption and environmental degradation. For example, construction industries are increasingly using recycled materials, such as reclaimed concrete and repurposed steel, to build infrastructure, thereby conserving natural resources and minimizing landfill waste. By demonstrating how waste can be transformed into valuable resources, GE not only mitigates environmental impact but also drives innovation and economic growth, highlighting its critical role in achieving a sustainable and resilient future.

1.9 Expected Impacts and Possible Outcomes of Implementing Green Engineering

To further enhance GE practices now and in the future, a multifaceted approach is essential. One significant avenue is the integration of cutting-edge technologies such as AI and the IoT within GE processes. These technologies facilitate smart automation and greater efficiency in resource management, which are critical for reducing waste and enhancing the sustainability of production processes. Researchers have demonstrated how AI can optimize building energy use, significantly reducing carbon footprints in urban developments [78–80]. Additionally, advancements in biotechnology, particularly in the synthesis of biodegradable materials, are pushing the boundaries of sustainable manufacturing by reducing reliance on non-renewable resources and minimizing ecological impacts. Another pivotal area for enhancement is education and training. By incorporating GE principles into the core curriculum of engineering courses and professional development programs, new engineers can be better equipped to design and implement sustainable solutions from the outset of their careers. Institutions like the Massachusetts Institute of Technology have integrated sustainability assessments and life cycle analysis training into their engineering programs, ensuring that graduates are proficient in these critical areas [81–84]. This educational emphasis not only fosters a new generation of engineers who prioritize sustainability but also ensures that these principles become standard practice across the engineering profession. Fostering collaboration across industries and disciplines can also accelerate the adoption and effectiveness of GE solutions. Interdisciplinary research teams can explore innovative approaches to complex environmental challenges by combining insights from material science, environmental science, and economic theory. Collaborative projects, such as the development of international standards for sustainable practices, are crucial for harmonizing efforts and maximizing the global impact of GE. By working together, industries and researchers can develop comprehensive solutions that address the multifaceted nature of sustainability issues, leading to more effective and widespread adoption of GE principles. The future of GE lies in the strategic integration of advanced technologies, the enhancement of educational programs, and the promotion of interdisciplinary collaboration. These efforts will ensure that GE continues to evolve and effectively address the pressing environmental challenges of our time, paving the way for a more sustainable and resilient future.

The continued enhancement of GE is anticipated to yield significant environmental, economic, and social benefits. Environmentally, GE promises a substantial reduction in waste and pollution, a decrease in natural resource depletion, and improved ecosystem health. These outcomes are critical for mitigating the impacts of climate change and supporting biodiversity. By adopting sustainable practices, industries can minimize their ecological footprint, contributing to the restoration and preservation of natural habitats and ecosystems. Economically, GE fosters resource efficiency, translating into cost savings for businesses through reduced material and energy expenditures and enhanced product life cycles. The principles of GE promote the use of renewable resources and more efficient manufacturing processes, which can lead to significant reductions in operational costs. Furthermore, the adoption of GE practices can open up new markets for green products and technologies, driving economic growth and innovation. Companies that invest in sustainable technologies and practices can gain a competitive advantage, attracting environmentally conscious consumers and investors. Socially, enhanced GE can lead to improved quality of life in various ways. Cleaner production techniques can reduce air and water pollution, leading to healthier communities. For example, industries that adopt GE principles are likely to produce fewer toxic emissions and pollutants, which can significantly improve public health outcomes. Additionally, GE can create more sustainable urban environments that offer residents better living conditions through green spaces and more efficient infrastructure. Cities that implement sustainable urban planning and development practices can provide their residents with healthier, more livable environments, featuring clean air, accessible green spaces, and energy-efficient buildings. Ultimately, the broad implementation of advanced GE practices promises to catalyze a transition to more sustainable industrial and urban systems. As noted in research, such transitions are essential not just for meeting specific environmental targets but also for ensuring long-term planetary health and human well-being. The role of GE in this transition is undoubtedly critical, as it provides the tools and methodologies necessary to design a more sustainable future [47, 85, 86]. By continuing to innovate and integrate GE principles into various sectors, society can move toward a more resilient and sustainable world, benefiting both current and future generations.

1.10 Conclusion

Throughout this review, we have explored the multifaceted nature of GE and its integral role within the framework of the CE. From the foundational principles that guide sustainable design to the practical applications seen in urban development and industrial processes, GE has proven to be a pivotal force in driving environmental sustainability across multiple sectors. The case studies of cities like Freiburg and companies like IKEA and Dell Technologies illustrate the successful implementation of these principles, showcasing significant advancements in sustainable practices that effectively minimize environmental impact while promoting economic viability. Freiburg, Germany, stands out as a model city for integrating GE into urban planning. Known as the "Green City," Freiburg has embraced ecological urban development since the 1970s, incorporating extensive public transportation networks, renewable energy initiatives, and sustainable building codes. The Vauban district exemplifies this approach with low-energy housing standards and a strong emphasis on communal green spaces. These efforts highlight how comprehensive planning strategies can reduce urban carbon footprints and enhance the quality of life for residents. Similarly, IKEA's commitment to sustainable practices demonstrates the effectiveness of CE principles in the retail sector. By designing products for longevity, ease of disassembly, and recycling, IKEA has established a closed-loop system that minimizes waste. Their initiatives to incorporate renewable materials and reduce carbon footprints further illustrate the company's dedication to sustainability. Dell Technologies also showcases the application of GE through its closed-loop recycling initiative and modular product designs, such as the OptiPlex 7070 Ultra. These efforts not only extend product lifespans but also reduce e-waste, setting industry standards for sustainable computing solutions. The synergies between GE and the CE have been highlighted, showing how interdisciplinary collaboration and the integration of innovative technologies can enhance resource efficiency and sustainability. Technologies such as AI, the IoT, and advanced robotics are transforming traditional production and consumption patterns, offering unprecedented opportunities for optimizing resource use and enhancing product lifecycles. Biotechnology and advancements in material science are also contributing to sustainable manufacturing by developing biodegradable materials and improving recycling processes. These interactions not only support environmental goals but also foster economic growth and social well-being, illustrating the comprehensive benefits of adopting GE practices. The ongoing evolution of GE is crucial for addressing the growing environmental challenges posed by modern industrial activities and urban development. Continued innovation in this field is essential for developing new materials, technologies, and processes that further reduce ecological footprints and promote the regeneration of natural systems. The importance of implementing these innovations cannot be overstressed, as it ensures that sustainable practices move from theoretical frameworks to real-world applications, leading to measurable environmental, economic, and social benefits. Moreover, the role of policy, education, and corporate responsibility in fostering the adoption of GE principles is paramount. Supportive regulatory frameworks and incentives are critical for encouraging industries and communities to invest in sustainable technologies and designs. Educational institutions play a vital role in equipping new engineers with the knowledge and skills necessary to implement sustainable solutions. Corporate responsibility and proactive sustainability strategies, as demonstrated by companies like IKEA and Dell, are essential for setting industry standards and influencing consumer behavior. This collective effort is essential for achieving global sustainability goals and transitioning toward a more sustainable future. The advancements discussed in this review underscore the potential of GE and the CE to create a resilient, sustainable world that benefits both people and the planet.

AI disclosure statement: The author acknowledges the use of AI-based tools for minor language editing or grammatical correction. However, all intellectual contributions, analyses, and conclusions presented in this chapter are entirely the author's own.

References

1. Dicks, A.P., *et al.*, A Systems Thinking Department: Fostering a Culture of Green Chemistry Practice among Students. *J. Chem. Educ.*, 96, 2836–2844, 2019.
2. Marteel-Parrish, A.E. and Abraham, M.A., Green chemistry and engineering: A pathway to sustainability, in: *Green Chem. Eng. A Pathw. to Sustain.*, pp. 1–361, 2013, 9780470413265.
3. Zaborniak, I., *et al.*, Controlled Polymer Synthesis Toward Green Chemistry: Deep Insights into Atom Transfer Radical Polymerization in Biobased Substitutes for Polar Aprotic Solvents. *ACS Sustain. Chem. Eng.*, 12, 4933–4945, 2024.
4. Anastas, P.T. and Zimmerman, J.B., Peer Reviewed: Design Through the 12 Principles of Green Engineering. *Environ. Sci. Technol.*, 37, 94A–101A, 2003.
5. Liu, Z., Adams, M., Cote, R.P., Geng, Y., Li, Y., Comparative study on the pathways of industrial parks towards sustainable development between China and Canada. *Resour. Conserv. Recycl.*, 128, 417–425, 2018.
6. Fairbrother, A., *et al.*, Toward Sustainable Environmental Quality: Priority Research Questions for North America. *Environ. Toxicol. Chem.*, 38, 1606, 2019.
7. Searchinger, T., *et al.*, Use of U.S. croplands for biofuels increases greenhouse gases through emissions from land-use change. *Science (80-.)*, 319, 1238–1240, 2008.
8. Allen, D.T. and Shonnard, D.R., Green engineering: Environmentally conscious design of chemical processes and products. *AlChE J.*, 47, 1906–1910, 2001.

9. Connelly, L. and Koshland, C.P., Exergy and industrial ecology. Part 2: A non-dimensional analysis of means to reduce resource depletion. *Exergy, Int. J.*, 1, 234–255, 2001.
10. Zhu, Q. and Sarkis, J., Relationships between operational practices and performance among early adopters of green supply chain management practices in Chinese manufacturing enterprises. *J. Oper. Manag.*, 22, 265–289, 2004.
11. UN Department of Economic and Social Affairs, World Urbanization Prospects. Demographic Research, vol. 12, 2018.
12. Nations, U., Revision of World Urbanization Prospects, United Nations, 2018.
13. Pongrácz, E., Re-defining the concepts of waste and waste management evolving the theory of waste management, 2002.
14. Sustainable Construction Overview, pp. 1–7, 2019.
15. Turner, B.L., *et al.*, Biodegradable elastomeric circuit boards from citric acid-based polyesters. *npj Flex. Electron.*, 7, 1, 1–14, 2023.
16. Laidler, K.J., Keith, J., Meiser, J.H., *Phys. Chem.*, 919, 1982.
17. van Bueren, E., van Bohemen, H., Itard, L., Visscher, H., *Sustainable urban environments: An ecosystem approach*, pp. 1–429, 2012, doi: 10.1007/978-94-007-1294-2.
18. Van Bueren, E., Chapter January 2012. 2014.
19. Silva, W.D.O., Fontana, M.E., de Almeida, B.M.J., Carmona Marques, P., Vidal, R., Analyzing the Attractiveness of Businesses to Receive Investments for a Creative and Innovative Transition to a Circular Economy: The Case of the Textile and Fashion Industry. *Sustainability*, 15, 2023.
20. Dhéret, B.C., Morosi, M., Frontini, A., Hedberg, A., Pardo, R., Towards a New Industrial Policy for Europe, European Policy Centre, 2014.
21. Velenturf, A.P.M. and Purnell, P., Principles for a sustainable circular economy. *Sustain. Prod. Consump.*, 27, 2021.
22. Bakker, C., Wang, F., Huisman, J., Den Hollander, M., Products that go round: Exploring product life extension through design. *J. Clean. Prod.*, 69, 2014.
23. Suchek, N., Fernandes, C.I., Kraus, S., Filser, M., Sjögrén, H., Innovation and the circular economy: A systematic literature review. *Bus. Strat. Environ.*, 30, 2021.
24. Ghisellini, P., Cialani, C., Ulgiati, S., A review on circular economy: The expected transition to a balanced interplay of environmental and economic systems. *J. Clean. Prod.*, 114, 2016.
25. Circular economy action plan - European Commission. https://environment.ec.europa.eu/strategy/circular-economy-action-plan_en.
26. Stahel, W.R., The circular economy. *Nature*, 531, 435–438, 2016.
27. Ellen MacArthur Foundation, Towards the circular economytransition. *J. Ind. Ecol.*, 2, 2013.
28. Yuan, Z., Bi, J., Moriguichi, Y., The circular economy: A new development strategy in China. *J. Ind. Ecol.*, 10, 4–8, 2006.

29. Lieder, M. and Rashid, A., Towards circular economy implementation: a comprehensive review in context of manufacturing industry. *J. Clean. Prod.*, 115, 36–51, 2016.
30. Tukker, A., Product services for a resource-efficient and circular economy - A review. *J. Clean. Prod.*, 97, 76–91, 2015.
31. Walters, G., Janzen, C., Maginnis, S., *Nature-based solutions to address global societal challenges*, 2016, doi: 10.2305/iucn.ch.2016.13.en.
32. Haase, D., *Integrating Ecosystem Services, Green Infrastructure and Nature-Based Solutions—New Perspectives in Sustainable Urban Land Management*, pp. 305–318, 2021, doi: 10.1007/978-3-030-50841-8_16.
33. Henriques, R., Figueiredo, F., Nunes, J., Product-Services for a Resource-Efficient and Circular Economy: An Updated Review. *Sustain.*, 15, 1–24, 2023.
34. Geissdoerfer, M., Savaget, P., Bocken, N.M.P., Hultink, E.J., The Circular Economy – A new sustainability paradigm? *J. Clean. Prod.*, 143, 757–768, 2017.
35. Hartley, K., Baldassarre, B., Kirchherr, J., Circular economy as crisis response: A primer. *J. Clean. Prod.*, 434, 140140, 2024.
36. Bocken, N.M.P., de Pauw, I., Bakker, C., van der Grinten, B., Product design and business model strategies for a circular economy. *J. Ind. Prod. Eng.*, 33, 308–320, 2016.
37. Wilber, D., Book Review: Cradle to Cradle: Remaking the Way We Make Things. William McDonough and Michael Braungart. 2002. North Point Press, New York. 208 pp. $27.50 paperback (978-0-86547-587-8). *Environ. Pract.*, 12, 92–94, 2010.
38. Yang, M., *et al.*, Circular economy strategies for combating climate change and other environmental issues. *Environ. Chem. Lett.*, 21, 2023.
39. Vladimirova, D., Waste to wealth: the circular economy advantage. *Prod. Plan. Control*, 28, 2017.
40. Morseletto, P., Mooren, C.E., Munaretto, S., Circular Economy of Water: Definition, Strategies and Challenges. *Circ. Econ. Sust.*, 2, 1463–1477, 2022. https://doi.org/10.1007/s43615-022-00165-x
41. Corvellec, H., Stowell, A.F., Johansson, N., Critiques of the circular economy. *J. Ind. Ecol.*, 26, 2022.
42. Prieto-Sandoval, V., Jaca, C., Ormazabal, M., Towards a consensus on the circular economy. *J. Clean. Prod.*, 179, 2018.
43. Grafström, J. and Aasma, S., Breaking circular economy barriers. *J. Clean. Prod.*, 292, 2021.
44. Kirchherr, J., Reike, D., Hekkert, M., Conceptualizing the circular economy: An analysis of 114 definitions. *Resour. Conserv. Recycl.*, 127, 221–232, 2017.
45. Cagno, E., Neri, A., Negri, M., Bassani, C.A., Lampertico, T., The role of digital technologies in operationalizing the circular economy transition: A systematic literature review. *Appl. Sci.*, 11, 2021.

46. Kirchherr, J., Yang, N.H.N., Schulze-Spüntrup, F., Heerink, M.J., Hartley, K., Conceptualizing the Circular Economy (Revisited): An Analysis of 221 Definitions. *Resour. Conserv. Recycl.*, 194, 107001, 2023.
47. Schröder, P., Anantharaman, M., Anggraeni, K., Foxon, T.J., *The Circular Economy and the Global South: Sustainable Lifestyles and Green Industrial Development*, 2019, doi: 10.4324/9780429434006.
48. Behún, M. and Behúnová, A., Advanced Innovation Technology of BIM in a Circular Economy. *Appl. Sci.*, 13, 2023.
49. Pieroni, M.P.P., McAloone, T.C., Pigosso, D.C.A., Business model innovation for circular economy and sustainability: A review of approaches. *J. Clean. Prod.*, 215, 2019.
50. Velzing, E.-J., Van der Meijden, A., Vreeswijk, K., Vrijhoef, R., Circularity in value chains for building materials, 2019, doi: 10.4995/carpe2019.2019.10196.
51. Averina, E., Frishammar, J., Parida, V., Assessing sustainability opportunities for circular business models. *Bus. Strat. Environ.*, 31, 2022.
52. Kirchherr, J., *et al.*, Barriers to the Circular Economy: Evidence From the European Union (EU). *Ecol. Econ.*, 150, 2018.
53. Milios, L., Overarching policy framework for product life extension in a circular economy—A bottom-up business perspective. *Environ. Policy Gov.*, 31, 2021.
54. Calisto Friant, M., Vermeulen, W.J.V., Salomone, R., Transition to a Sustainable Circular Society: More than Just Resource Efficiency. *Circ. Econ. Sustain.*, 4, 2024.
55. Webster, K., *The Circular Economy: A Wealth of Flows*, vol. 2, The Ellen Macarthur Foundation, 2015.
56. Martínez León, H.C., Bridging theory and practice with Lean Six Sigma capstone design projects. *Qual. Assur. Educ.*, 27, 2019.
57. Harala, L., Alkki, L., Aarikka-Stenroos, L., Al-Najjar, A., Malmqvist, T., Industrial ecosystem renewal towards circularity to achieve the benefits of reuse - Learning from circular construction. *J. Clean. Prod.*, 389, 2023.
58. Bressanelli, G., Adrodegari, F., Perona, M., Saccani, N., Exploring how usage-focused business models enable circular economy through digital technologies. *Sustain.*, 10, 2018.
59. Jiménez Estévez, P., *et al.*, The Circular Economy and its effects on reducing the carbon footprint in IKEA company. *Int. J. Bus. Res. Manag.*, 2022.
60. Coppola, C., Vollero, A., Siano, A., Developing dynamic capabilities for the circular economy in the textile and clothing industry in Italy: A natural-resource-based view. *Bus. Strat. Environ.*, 32, 2023.
61. McDonough, W., Braungart, M., Anastas, P.T., Zimmerman, J.B., Applying the Principles of Green Engineering to Cradle-to-Cradle Design. *Environ. Sci. Technol.*, 37, 2003.
62. Newman, P., Beatley, T., Boyer, H., *Resilient Cities: Overcoming Fossil Fuel Dependence*, 2017, doi: 10.5822/978-1-61091- 686-8.

63. Allan, A., Resilient Cities: Overcoming Fossil Fuel Dependence. *Urban Pol. Res.*, 38, 2020.
64. Boz, Z., Korhonen, V., Sand, C.K., Consumer considerations for the implementation of sustainable packaging: A review. *Sustain.*, 12, 2020.
65. Parolin, G., McAloone, T.C., Pigosso, D.C.A., How can technology assessment tools support sustainable innovation? A systematic literature review and synthesis. *Technovation*, 129, 2024.
66. Bennett, J.W., Pearce, D.W., Turner, R.K., Economics of Natural Resources and the Environment. Baltimore MD: Johns Hopkins University Press, 1990, 378 pp.50 paper. *Am. J. Agric. Econ.*, 73, 227–228, 1991.
67. Vilochani, S., McAloone, T.C., Pigosso, D.C.A., Integration of sustainability into product development: insights from an industry survey. *Proc. Des. Soc.*, 4, 1517–1526, 2024.
68. Rossi, L., Leone, D., Barni, A., Fontana, A., Assessing the Sustainability of Industrial Equipment Life Extension Strategies through a Life Cycle Approach: Methodology and Practical Guidelines. *Processes*, 10, 2022.
69. Perotti, F.A., Bargoni, A., De Bernardi, P., Rozsa, Z., Fostering circular economy through open innovation: Insights from multiple case study. *Bus. Ethics Environ. Responsib.*, 2024, doi: 10.1111/beer.12657.
70. Jones, P., Comfort, D., Hillier, D., Sustainability in the global shop window. *Int. J. Retail Distrib. Manage.*, 39, 2011.
71. Meade, L. and Presley, A., Building the business case for sustainable supply chains, in: *Handbook on the Sustainable Supply Chain*, 2019.
72. Freytag, T., Gössling, S., Mössner, S., Living the green city: Freiburg's Solarsiedlung between narratives and practices of sustainable urban development. *Local Environ.*, 19, 2014.
73. Adem Esmail, B., *et al.*, Greening cities through urban planning: A literature review on the uptake of concepts and methods in Stockholm. *Urban For. Urban Green.*, 72, 2022.
74. Blevis, E., Sustainable interaction design: Invention & disposal, renewal & reuse, in: *Conference on Human Factors in Computing Systems - Proceedings*, 2007, doi: 10.1145/1240624.1240705.
75. Koch, L., Roberts, S., Ellis, J.E., Dell: The business case for a sustainable supply chain, in: *Putting Purpose Into Practice: The Economics of Mutuality*, 2021.
76. Gayathri Ganesan, N., Samal, S., Kannan, V., Kumar Rathnasamy, S., Rangarajan, V., Chapter 12 Bacillus lipopeptide-based antifungal agents for plant disease control, in: *Plant Prot.*, pp. 313–336, 2022.
77. Tiwari, U., Ganesan, N.G., Junnarkar, J., Rangarajan, V., Toward the formulation of bio-cosmetic nanoemulsions : from plant-derived to microbial-derived ingredients. 2021.
78. Onwusinkwue, S., *et al.*, Artificial intelligence (AI) in renewable energy: A review of predictive maintenance and energy optimization. *World J. Adv. Res. Rev.*, 21, 2024.

79. Ohalete, N.C., *et al.*, AI-driven solutions in renewable energy: A review of data science applications in solar and wind energy optimization. *World J. Adv. Res. Rev.*, 20, 2023.
80. Fan, Z., Yan, Z., Wen, S., Deep Learning and Artificial Intelligence in Sustainability: A Review of SDGs, Renewable Energy, and Environmental Health. *Sustainability*, 15, 2023.
81. Okada, Y., Kishita, Y., Nomaguchi, Y., Yano, T., Ohtomi, K., Backcasting-Based Method for Designing Roadmaps to Achieve a Sustainable Future. *IEEE Trans. Eng. Manag.*, 69, 2022.
82. Claessens, F., *et al.*, A co-creation roadmap towards sustainable quality of care: A multi- method study. *PLoS One*, 17, 2022.
83. Faludi, J., *et al.*, A research roadmap for sustainable design methods and tools. *Sustainability*, 12, 2020.
84. IEA, Technology Roadmap: Delivering Sustainable Bioenergy, IEA Publ., 2017.
85. Azizuddin, M., Shamsuzzoha, A., Piya, S., Influence of circular economy phenomenon to fulfil global sustainable development goal: Perspective from Bangladesh. *Sustainability*, 13, 2021.
86. Schroeder, P., Anggraeni, K., Weber, U., The Relevance of Circular Economy Practices to the Sustainable Development Goals. *J. Ind. Ecol.*, 23, 2019.

2
Green Metrics

Gracia Saral Gladison, Ramya Muniasamy and Senthilkumar Rathnasamy*

Green Separation Engineering Laboratory, School of Chemical & Biotechnology, SASTRA Deemed to be University, Thanjavur, Tamilnadu, India

Abstract
This chapter details the diverse metrics used for quantifying the greenness of a process or a product. While various green approaches and their multifaceted demand have been detailed in the previous chapter, their quantification is essential to reflect the significance of their widespread adaptation. This has been satisfied by enlisting the 12 principles of greenness that elaborate the approach for sustainability quantification. Further, several scalar equations like atom economy, reaction mass index, E-factor, and mass intensity have also been introduced which aid in disseminating these principles. Additionally, critical case studies that disseminate these green principles are also being discussed. Finally, the scope of Sustainable Development Goals proposed by the United Nations to achieve a green index is discussed, along with its prospects and challenges. This chapter is intended to provide a detailed overview of the recent development in a global quantification approach for measuring sustainability.

Keywords: Green chemistry, sustainable development goals, atom economy, reaction mass efficiency, eco-scale, environmental protection, renewable resources, public awareness

2.1 Introduction to Green Chemistry—Demand and Scope

Green chemistry is one of the most influential ideas in the physical sciences. When Paul Anastas of the US Environmental Protection Agency (EPA) helped create the Green Chemistry Program in 1991, the phrase

**Corresponding author*: senthilrathna@sastra.ac.in

Senthilkumar Rathnasamy, Vivek Rangarajan and Umile Gianfranco Spizzirri (eds.) Green Engineering and Sustainable Technology: Tools and Techniques, (45–70) © 2026 Scrivener Publishing LLC

was initially used. The main goal of this initiative was "to promote the research, development, and implementation of innovative chemical technologies that accomplish pollution prevention in a both scientifically and cost-effective manner." The publication of two significant papers on green metrics [1, 2] influenced the development of the 12 Principles of Green Chemistry in 1998. Further propelled, this environmental friendly strategy aligns with the principles of green chemistry by minimizing waste, reducing the use of hazardous substances, and promoting sustainable resource utilization. Since then, green chemistry has gained popularity and has been included in thousands of academic papers, many books on basic and contemporary research, as well as numerous educational works that aim to include the subject in a university-level course or curriculum. These methods aim to challenge the long-held belief that a synthetic product yield may be used to accurately represent process efficiency. Rather, by approaching synthetic efficiency from a wider angle, researchers are empowered to improve the efficiency and sustainability of the systems they have built.

Demand and Scope for Green Chemistry

Globally, government laws pertaining to environmental sustainability are becoming more stringent, compelling businesses to adopt green measures in order to comply. The European Union's Packaging and Packaging Waste Regulation requires effective sustainability practices measurement, which sets extremely high recyclability standards by 2030 [3]. Corporate Responsibility: A growing number of investors and customers hold businesses responsible for their environmental effects. The majority of businesses today disclose their sustainability initiatives to the public, with 86% acknowledging their susceptibility to climate-related hazards. This necessitates a dedication to sustainability and the use of reliable green measures to gauge progress. Investment Trends: Investors are focusing more on environmental, social, and governance (ESG) criteria in investment decisions. The sustainability metrics allow for companies to report and present their environmental performance transparently, hence attracting investment. Technological Advancements: Technologies like Internet of Things and data analytics make it possible to monitor sustainability metrics in real time, enabling organizations to better track their environmental performance.

As more businesses and organizations realize how important sustainability is to their operations, the potential for green metrics is expanding significantly. These measures offer crucial instruments for evaluating environmental performance, directing choices, and stimulating creativity, particularly in the manufacturing and chemistry industries. When assessing the sustainability of chemical processes, green indicators are crucial. Waste generation and resource efficiency in chemical synthesis are evaluated with the use of metrics like atom economy, E-factor, and process mass intensity (PMI). By using these indicators, businesses can find ways to lessen their environmental effect without sacrificing productivity. Pharmaceutical sector: As it measures the environmental impact of drug production, the pharmaceutical sector is adopting more environmentally friendly metrics [4]. One such statistic is Manufacturing Mass Intensity, which incorporates cleaning and recycling efforts in addition to the mass of active medicinal substances to measure the wider effects of production operations. Because of increasing public attention and regulatory demands, sustainability is crucial.

2.2 Green Metrics—Pillars of Green Chemistry

The study of "green chemistry" focuses on creating chemical processes and products that use fewer or no extremely dangerous ingredients. Below Table 2.1 shows 12 principle of green chemistry along with its description serve as the foundation for more environmentally friendly chemical operations, which lower environmental impact and guarantee increased safety and efficiency in the production of chemicals [5].

Table 2.1 Principles of green chemistry [5].

Sl. no.	Principles	Description
1)	Waste prevention	Waste prevention is superior to waste treatment, according to the first and most significant concept.
2)	The economy of atoms	It is recommended that chemists create synthetic processes that optimize the amount of the raw ingredients incorporated into the finished product.
3)	Safer synthesis	From the perspective of both individuals and the environment, it is crucial to design chemical processes that produce and use fewer amounts of harmful compounds whenever possible.

(*Continued*)

Table 2.1 Principles of green chemistry [5]. (*Continued*)

Sl. no.	Principles	Description
4)	Safer chemicals	Chemical products ought to be made with the least amount of toxicity possible while yet fulfilling their intended purpose.
5)	Safer auxiliaries and solvents	Solvents and separation agents are examples of auxiliary compounds that should be eliminated wherever possible and made harmless when used.
6)	Efficiency in energy use	The economic and environmental effects of energy use in chemical synthesis ought to be detectable and controllable. Whenever possible, procedures that are conducted at room temperature and pressure should be employed.
7)	Renewable feedstocks	When technically and financially possible, starting materials derived from renewable feedstocks should be used.
8)	Reduce derivatives	For the objective of minimizing waste, blocking groups, protection/deprotection, and temporary changes to physical/chemical processes should be avoided.
9)	Catalysis	Catalytic reagents with selective designs are more effective and generally produce less waste than stoichiometric reagents.
10)	Create for deterioration	Chemical products ought to be made in a way that keeps them from building up in the environment.
11)	Analysis in real time	Establishing and putting into practice analytical techniques that provide real-time, in-process monitoring and control before hazardous compounds are created is essential.
12)	Safer chemicals to prevent mishaps	By choosing naturally safer materials, the risk of chemical mishaps like leaks, explosions, and flames should be decreased.

The discipline's objective was stated as "the design of chemical products and processes that reduce or eliminate the use or generation of hazardous substances" [6] in the EPA's Green Chemistry Program Fact Sheet from 1991. To assist others in achieving this objective, Anastas and Warner developed 12 essential principles in 1998 [7].

2.2.1 Atom Economy and Reaction Mass Efficiency

One of the core ideas of green chemistry is atom economy, which measures how well the atoms of the reactants are employed to create the intended products in chemical processes. Barry Trost coined the phrase in 1991 [1] to evaluate the sustainability of chemical processes while keeping molecular waste reduction in mind [8]. The mathematical definition of "atom economy" is as follows:

$$Atom\ Economy = \left[\frac{Molecular\ weight\ of\ desired\ product}{Total\ molecular\ weight\ of\ all\ reactants}\right] * 100\%$$

This formula determines the percentage of atoms from the reactants included in the finished product. When every atom in the reactants is converted into the intended product and not wasted, the reaction is said to have 100% atom economy.

2.2.1.1 The Atom Economy's Significance

- Waste Minimization: Atom economy encourages chemists to create processes that optimize the incorporation of reactant atoms into products, so directly addressing waste creation. This idea aligns with the goal of green chemistry, which is to stop pollution at the molecular level.
- Economic Efficiency: Because there is less waste of the initial material, high atom economy is frequently associated with cheaper material costs. This financial benefit incentivizes businesses to adopt more environmentally friendly practices and could lead to significant cost reductions, particularly at the industrial level.
- Environmental Impact: The high atom economy can have a lesser environmental impact because it produces fewer trash and toxic byproducts. This is particularly important for businesses where disposing of waste can be costly and harmful to the environment.

2.2.1.2 Atom Economy Examples

- Ibuprofen Synthesis: A comparison of two synthetic methods of synthesis of ibuprofen using two distinct methods is an excellent example of the atom economy notion. With an atom economy of roughly 40%, the older technique wastes 60% of the reagents; in contrast, the newer route has an atom economy of 77%, which significantly reduces waste and makes it more sustainable.
- The combustion of methane using an atom economy can be used to evaluate the methane combustion reaction, which produces carbon dioxide and water when methane and oxygen combine. Therefore, based on molecular weights concerning the molecules consumed in the process, both products provide a basis for calculating the atom economy.

Although it is preferable to maximize atom economy, other crucial aspects like yield, purity, and ease of product isolation should not be sacrificed in the process. For instance, side reactions or trouble isolating the target product can result in a reaction with a high atom economy but a low yield. Therefore, while designing chemical processes, scientists need to take these factors into account holistically.

An essential tenet of green chemistry is the atom economy, which encourages sustainable processes by minimizing waste production and making effective use of basic resources. Chemists can help create more ecologically friendly chemical processes that support larger sustainability objectives by concentrating on this parameter. Greener chemistry solutions could be significantly advanced by incorporating atom economy into chemical education and industry processes.

One of the key indicators in green chemistry is reaction mass efficiency (RME), which quantifies the effectiveness of a chemical reaction by indicating how well the mass of reactants is converted into the intended product [9]. The metric RME was examined with an emphasis on reactants and reaction efficiency. It enables us to consider a reaction holistically, [10] taking into account the yield, stoichiometry, and total mass of components involved, more so than other metrics like atom economy. By definition, RME is as follows:

For a general response, X + Y = Z

$$\text{RME} = \left[\frac{\textit{Molecular weight of product } Z}{\textit{Molecular weight of } X + (\textit{molecular weight of } Y * \textit{molar ratio} \frac{Y}{X})}\right] * \textit{yield}$$

Or simply as

$$\text{RME}=\left[\frac{Mass\ of\ product\ Z}{mass\ of\ X+mass\ of\ Y}\right]*100$$

This formula considers all of the reactants, solvents, and other components involved in the reaction, as well as the mass of the final product [11].

RME can be viewed as a productivity metric that goes beyond waste production. Elements of the RME Yield:

- A reaction's mass of recovered product in relation to its theoretical maximum.
- Yield is crucial because, even in reactions with great atom economy, low yield can result in the production of significant amounts of waste.
- The efficiency with which reactants are transformed into products is measured by the atom economy.
- RME incorporates yield and stoichiometry with conversion efficiency, whereas atom economy alone stresses this element.
- All surplus reactants that may have been employed in a reaction are taken into account by the stoichiometric factor.
- The overall efficiency is impacted, for example, when more than one reactant is required to propel the reaction to completion.
- Recovery Parameters for Materials: RME takes into account the process' ability to recover solvents and other materials.

2.2.1.3 *Significance of Reaction Mass Efficiency*

- Holistic Assessment: By taking into account all of the elements involved, such as catalysts and solvents, RME offers a more comprehensive picture of a reaction's sustainability. This all-encompassing method aids in pinpointing places where chemical processes need to be improved [11].
- Leading Sustainable Practices: Chemists use RME-based reaction design to maximize product yields with the least amount of waste and resources. This aligns with green chemistry concepts like reducing environmental impact.

- Industrial Significance: RME faithfully replicates real-world circumstances that represent the influence of multiple variables on efficiency, it is extremely important in industrial applications. By highlighting inefficiencies in traditional ways, it forces industries to adopt more environmentally friendly practices.

2.2.1.4 *Examples of RME*

- Esterification Reaction: Determine how n-butanol and acetic acid esterify to produce n-butyl acetate. To find out how efficiently these materials have been used, compute RME if the intended product yield is 40 g from 37 g of non-benign reagents.
- Catalytic Mechanisms: RME calculations in heterogeneous catalysis can reveal significant variations in process efficiency between laboratory-scale and industrial-scale operations. For instance, the nitration process used to create aniline from benzene may result in greater RME values than a less effective laboratory method.
- Measurement Complexity: Detailed information on all materials involved in a reaction, including by-products and losses during purification, is necessary for an accurate measurement of RME. Such intricacy may prevent consistent use in different reactions.
- Emphasis on Non-Benign Materials: While RME takes into account all materials used, it is important to highlight how environmentally acceptable the materials are. RME can be enhanced by metrics that concentrate on non-benign reagents, such as effective mass yield (EMY). RME, which combines yield, atom economy, and material utilization to provide an overall metric that improves the sustainability of chemical processes, is a crucial yardstick in green chemistry. Prioritizing RME in industrial and chemical education can assist scientists in developing more environmentally friendly manufacturing processes.

2.2.2 Process Mass Intensity and E-Factor

- Two essential metrics in green chemistry for evaluating the sustainability and environmental impact of chemical processes [12] are PMI and E-factor. Both measures provide information about how resources are used, waste is produced, and how efficiently chemicals are manufactured overall, particularly in the pharmaceutical sector [13].

- Although it has been used widely within SmithKline Beecham for almost four years before being documented in a publication, PMI was first published in 2001. [14] PMI was an effort to draw attention to the inefficiency of a common pharmaceutical process, which was not a topic of significant interest in the pharmaceutical business at the time. PMI took into consideration all reactants, reagents, catalysts, solvents, and any additional materials used in product work-up and isolation. It was based on RME. Calculating this is not too hard.
    - $\text{Mass Intensity [MI]} = \dfrac{\textit{Total mass in reaction chamber } [kg]}{\textit{Mass of product } [kg]}$
- The total mass of all the elements necessary to create a given quantity of product—raw materials, reactants, solvents, and catalysts—is known as the PMI [15]. The formula below can be used to compute it:
    - $\text{PMI} = \dfrac{\textit{Mass of raw materials input}}{\textit{Mass of product}}$
- This metric helps determine the "greenness" of a chemical reaction because it will provide a thorough understanding of the resources used during the synthesis process. The significance of the PMI holistic evaluation All of the components used in a process are taken into consideration by PMI, including work-up chemicals and solvents, which are often major sources of waste. This all-encompassing method aids in detecting chemical process inefficiencies.
- Chemists can identify even more sustainable alternatives by examining various processes and their corresponding PMI values. Waste minimization and resource optimization are the core aspects of green chemistry.
- Instruments for Enhancement: A number of tools and calculators have been created to help chemists evaluate and compare various synthetic pathways more quickly by simplifying PMI assessments. To improve the efficiency of this

procedure, the ACS GCI Pharmaceutical Roundtable, for instance, has created a number of PMI calculators.

- Issues with PMI Misinterpretation: If yield and atom economy are not taken into account, then PMI might be deceptive. For instance, if high-quality goods are produced, then reactions with a high PMI may still be effective. Impact of Solvents: In some situations, the solvents utilized account for as much as 70% of the PMI value. A situation like that could distort how efficient we think a procedure is. Therefore, choosing the right solvent is crucial for increasing greenness overall.
- Another green chemistry performance indicator that quantifies the quantity of waste generated per unit of product produced is the E-factor [16]. This can be computed as follows:
  - $\text{E-factor} = \dfrac{\textit{Mass of waste}}{\textit{Mass of product}}$
- Because it concentrates on waste output rather than just material input, the E-factor aids in understanding the environmental effects of a chemical process.

The significance of focusing on e-factor waste management:

- The E-factor highlights how important it is to reduce waste in chemical operations. The procedure is more environmentally friendly and sustainable when the E-factor is lower.
- Comparison of industry to industry:
  The E-factor makes it easier to compare different industries and identify best practices for waste reduction in fine chemicals, pharmaceuticals, and other chemical process operations. Chemical engineers may design more efficient processes that align with sustainability goals by using the E-factor, which clearly identifies the waste generation measure.

Two important metrics in green chemistry that offer important information about the sustainability of chemical processes are [17] PMI and E-factor. The E-factor stresses waste creation in relation to product output, whereas PMI concentrates on resource use by taking into account all materials involved in synthesis. When combined, these indicators assist industries and chemists in pinpointing areas that require improvement, streamlining procedures, and eventually promoting more environmentally friendly chemical manufacturing methods [18]. Organizations can enhance their dedication to environmental stewardship and advance more general sustainability goals by incorporating these measures into their research and development efforts.

2.2.3 Eco-Scale and Waste Reduction Algorithm

Two essential instruments in green chemistry for evaluating and enhancing the sustainability of chemical processes are the Eco-Scale and Waste Reduction Algorithm. Waste reduction, resource use, and the overall environmental impact of chemical reactions are the main focus of the measurements. A semi-quantitative method for evaluating synthetic reactions' environmental performance is the Eco-Scale. It provides a score ranging from 0 to 100, representing the optimal reaction in terms of high yield, low cost, low danger, and low environmental effect [19, 20].

The total Eco-Scale score was calculated as follows:

Eco-Scale = 100 – sum of individual penalties

Comprehensive Evaluation: The Eco-Scale takes into account a number of factors, such as the following:

- Cost of reagents
- Yield
- Safety issues
- Energy consumption
- Work-up
- Purification

User-Friendly Style: The Eco-Scale is clear, easy to use, and available to chemists working in academia or commercial research settings.
Benefits of the Eco-Scale Holistic Evaluation: The Eco-Scale provides a comprehensive assessment of a reaction's greenness by taking into account a number of sustainability-related aspects.

Improvement Criterion: By highlighting opportunities for improvement in chemists' synthesis techniques, the grading system promotes cleaner practice.
Comparative Criteria: The Eco-Scale makes it simple to compare various synthetic approaches or processes [21].

- Restrictions on the Eco-Scale
- Absence of detail regarding the risks: It does not identify the precise types of risks connected to the materials used, even if its score is determined by the penalties.
- Subjectivity in Cost Evaluation: Reagents' "cheap" status will vary over time and be mostly determined by availability.

2.2.3.1 Reduction of Waste Algorithm

Metrics like the Eco-Scale, which concentrate on reducing waste generated in chemical processes, are enhanced by the Waste Reduction Algorithm [22]. By locating waste sources and suggesting adjustments to reduce waste output, this algorithm is intended to optimize processes.

Key Components of the Waste Reduction Method

- Identification of Waste: The program begins by identifying every kind of waste generated during a chemical process, such as the following:
- Unreacted initial components
- Byproducts Materials
- Auxiliary-solvents
- Waste Quantification: This indicates how much waste is generated for every mass of product. Calculations akin to those required for E-Factor analysis can thus be performed.
- Process Optimization: To reduce the quantity of waste generated, the algorithm may suggest the following modifications:
- Reaction condition optimization (temperature and pressure)
- Making use of more potent catalysts
- Techniques for recycling reagents and solvents

To provide a comprehensive understanding of sustainability in chemical processes, the Waste Reduction Algorithm can be combined with other green metrics as the E-Factor and PMI. In conclusion, the Eco-Scale and

Waste Reduction Algorithm are crucial instruments that enable the evaluation and improvement of chemical reactions. The tools help chemists create more ecologically friendly synthetic processes by emphasizing sustainability parameters including yield, cost, safety, and waste generation. By assuring safer procedures in labs and industry, implementation can lead to significant advancements in lowering the environmental effect of chemical manufacture.

2.2.4 Case Studies—Role of Green Metrics in Industrial Sustainability

Green metrics, which offer frameworks for measuring and improving environmental performance, are among the most significant elements for industrial sustainability. The majority of businesses in a variety of sectors have adopted these measurements in order to streamline operations, cut waste, and support sustainability objectives. The following noteworthy case studies demonstrate how green metrics are applied in industrial contexts.

1. Sunfit Rooftop Solar Initiative by Tata Power
Tata Power's rooftop solar projects in 90 Indian cities generated 421 million watts of electricity. It makes the most contribution to the sustainability of renewable energy sources by using vacant rooftops to generate green electricity. It illustrates how green metrics may be used practically to assess energy efficiency as well as its impact on sustainability.

2. UPS ORION: Effective Transportation
Because transportation is responsible for around 30% of greenhouse gas emissions in the United States, UPS created ORION, an AI-based route optimization system, to increase transportation efficiency [23]. UPS has decreased pollutants and increased fuel efficiency by streamlining delivery routes and cutting down on pointless turns. The significance of operational indicators in evaluating and enhancing sustainability in the logistics and transportation industries is shown by this case.

3. Green Building Practices at Swire Properties
One Taikoo Place is a green building constructed by Swire Properties with the goal of reducing greenhouse gas emissions. In order to enhance energy efficiency and smart lighting systems that adapt to the availability of natural light, the project made use of 3D modeling techniques. Additionally, Swire put in place a biodiesel production plant that turns used cooking oil into fuel. This example shows how using green construction measures and standards can result in notable carbon footprint reductions.

4. H&M: Initiatives for the Circular Economy
In an effort to promote recycling and reusing, H&M launched the "Let's Close the Gap" initiative, which gathers used clothing from consumers. By turning garbage into raw materials for new products, this program embodies the ideas of a circular economy [24]. H&M lowers resource usage and improves its overall environmental performance by measuring the effects of these actions using sustainability indicators.

5. Danfoss: A Path for Carbon Reduction
To decarbonize supply chains in industries with high emissions, Danfoss and McKinsey worked together to develop a comprehensive carbon reduction roadmap. In order to show how businesses can utilize sustainability indicators to create meaningful environmental changes, our effort involves setting quantifiable targets for emissions reductions and methods to meet those targets.

6. The Path to GreenCo Certification for Tata Steel
With an emphasis on sustainable practices throughout all operations, Tata Steel's processing and distribution facility in Pune started the GreenCo certification process. It received recognized for its efforts to promote industrial sustainability by implementing green measures in its operations related to waste management, resource usage, and energy efficiency.

7. Digital Wind Farms by General Electric
Digital twin technology was used by General Electric to maximize the operation of their wind turbines, enabling *in situ* real-time monitoring and real-time performance modifications. In certain instances, the result was an extra 10% of energy from the wind farms, demonstrating how the use of digital technology may improve sustainability indicators.

These case studies demonstrate how green metrics are used in a variety of industries and how they may be used to promote sustainability and lessen environmental impact. By employing these indicators, businesses will enhance their operational effectiveness and support global sustainability objectives. Therefore, including green metrics into corporate plans is a means of promoting creativity, improving resource management, and achieving long-term sustainability goals.

2.3 Sustainable Development Goals—Foundations of Green Society

The United Nations established the 17 Sustainable Development Goals (SDGs) in 2015 as a component of the 2030 Agenda for Sustainable

Development [25]. Figure 2.1 shows the list of sustainable development goals. Poverty, inequality, climate change, environmental degradation, peace, and justice are a few of the major issues that are addressed. By 2030, everyone should have a sustainable future. The list of goals is as follows [20]:

- Eliminate all types of poverty worldwide. There is no poverty.
- Zero Hunger aims to end hunger: Increase nutrition and food security, and advance sustainable agriculture.
- Good Health and Well-Being: Everyone, regardless of age, should lead healthy lives and be happy.
- Quality Education: Make sure that everyone has access to inclusive, egalitarian, high-quality education, and encourage possibilities for lifelong learning.
- Gender Equality: Give girls and boys the same chances, and give women more influence.
- Water and Sanitation: Make sure that everyone has access to and can sustainably manage water and sanitation.
- Clean and Inexpensive Energy: Make sure everyone has access to modern, sustainable, reasonably priced, and safe energy.
- Decent Work and Economic Growth: Maintain full and productive employment, decent work for all, and inclusive and sustainable economic growth. Infrastructure, innovation, and

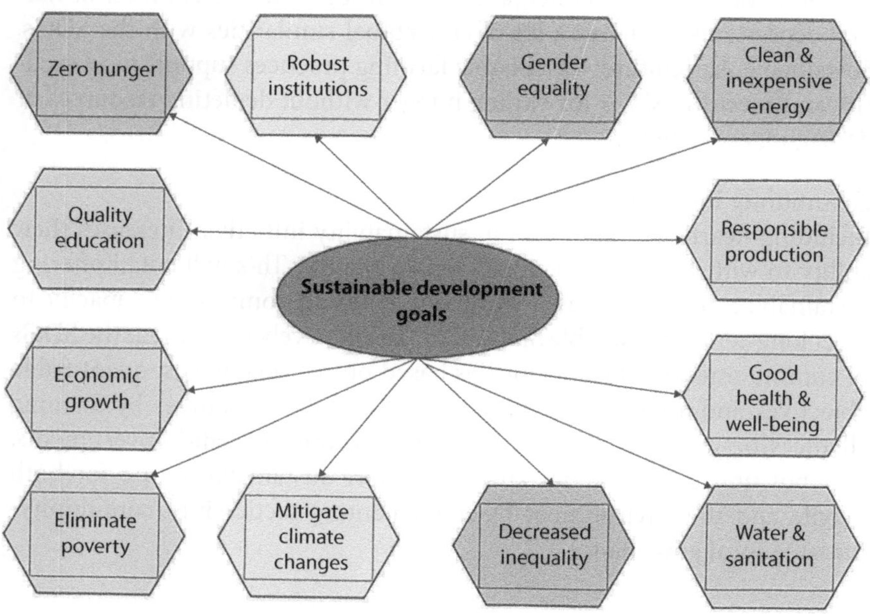

Figure 2.1 Sustainable development goals.

industry encourage innovation, encourage inclusive and sustainable industrialization, and provide robust infrastructure.
- Decreased Inequality: Decrease inequality both within and between nations, neighborhoods, and cities that are sustainable. Create inclusive, secure, resilient, and sustainable cities and human settlements.
- Responsible Production and Consumption: To guarantee sustainable patterns of production and consumption.
- Take action to prevent climate change and its consequences: Utilize the seas, oceans, and marine resources responsibly to support sustainable development while preserving submerged life. Controlling forestry operations, preventing landslides, and eliminating desertification.
- Peace, Justice, and Robust Institutions: Encourage inclusive and peaceful societies for long-term growth, ensure that everyone has access to justice, and establish strong, accountable institutions at all levels.
- Collaborations for the Objectives: Boost implementation tools and revitalize the international collaboration for sustainable development.

A green society emphasizes sustainability by ensuring biodiversity conservation and environmental health through the preservation of natural ecosystems; it shares a lot of conceptual similarities with the SDGs. Sustainable Agriculture: Sustainable farming practices support food security and meet the SDGs for ending hunger without depleting resources or damaging ecosystems.

Community Involvement:
Including nearby communities in sustainability initiatives increases their ability to withstand the effects of climate change. This will entail offering sustainability education as a means of building community capacity to meet long-term, sustainable SDGs [26]. To effectively implement the SDGs in communities, establish priorities based on regional needs, quantifiable objectives and associated particular SDGs. Utilize resources by forming alliances between communities, NGOs, corporations, and governments. Last but not least, societies aim for a more sustainable future for both people and the environment by implementing tactics into community-initiated local governance.

2.3.1 Prospects and Challenges of SDGs

Because green chemistry encourages ecologically beneficial methods of producing and using chemicals, it is a field that is closely related to the growth of SDGs [27]. The opportunities and difficulties of integrating green chemistry into the SDGs' framework are highlighted in this overview [28].

2.3.1.1 Prospects of SDGs

- At the source, Reduce and Recycle: Green chemistry takes into account stopping pollution and waste at its source. "Responsible Consumption and Production" follows immediately from this. Green chemistry significantly lowers the environmental impact of chemical manufacturing by designing processes that use fewer emissions and fewer dangerous chemicals.
- Utilizing Renewable Resources: The switch from fossil fuels to renewable resources, including biomass, aligns with several SDGs, including Climate Action and Affordable and Clean Energy. This shift helps to mitigate climate change by lowering carbon emissions and decreasing dependency on non-renewable resources.
- Chemical Process Innovation: Green chemistry advancements can provide new, more sustainable and effective materials and procedures. Innovations such as biocatalysis, for instance, can lower energy usage and boost industrial efficiency. By encouraging sustainable industrial practices, this aligns with industry, innovation, and infrastructure.
- Green chemistry requires economic growth through production process cost reductions and the development of new markets for environmentally friendly products in order to maintain growth. Generally speaking, increased efficiency tends to lower operating costs for both consumers and corporations.
- Healthier Communities: The creation of safer chemical goods promotes good health and well-being by improving public health. In order to provide healthier living conditions, green chemistry aims to create chemicals that are less harmful to both humans and the environment.

2.3.1.2 Challenges of SDGs

- Funding and Investment Gaps: Green chemistry projects frequently encounter financial limitations in spite of their potential. Innovation and the expansion of sustainable technologies—both essential for accomplishing the SDGs—are hampered by a lack of funding for research and development.
- Regulatory Obstacles: Adoption of green chemical processes is not entirely supported or encouraged by current regulations. When firms want to adopt greener processes in compliance with different rules, the absence of standardized standards poses regulatory obstacles.
- Public Education and Awareness: Stakeholders, including legislators, business executives, and the general public, need to be better informed about the advantages of green chemistry. To encourage awareness and adoption of sustainable practices in the chemical sector, education programs are crucial.
- Technological Limitations: While some green chemistry advancements appear to be very promising, many of these technologies are still in the early phases of development or are not yet widely applicable. To make widespread use of technology, these obstacles must be removed.
- Global Inequalities: Depending on variations in infrastructure, educational attainment, and available resources, the application of green chemistry practices varies significantly by area. Applying the SDGs equitably is necessary to address inequities as a global issue and advance the goals.

2.3.2 Economic Aspects of SDGs

By encouraging economic growth without creating or worsening adverse environmental effects, green chemistry helps to accomplish the SDGs. In order to illustrate the advantages and difficulties of green chemistry, this analysis focuses on its economics in relation to the SDGs.

- Cost Savings: The manufacturing of chemicals can be greatly reduced by applying the concepts of green chemistry. Costs associated with disposal, treatment, and storage can be decreased by reducing waste and hazardous items.

By encouraging sustainable manufacturing techniques, this efficiency boosts profitability and fosters responsible consumption and production.

- Innovation and Market Growth: By developing new, more sustainable products and processes, green chemistry stimulates innovation. This invention has the potential to boost the economy by creating new markets for eco-friendly goods. For instance, new sustainability-focused enterprises may arise as a result of the discovery of biodegradable materials or renewable energy sources.
- Employment Creation: The shift to a green economy encourages employment in sustainable agriculture, waste management, and renewable energy. Employment opportunities are being created as a result of industries adopting greener practices, which do call for a staff with more advanced technologies and procedures.
- Resource Efficiency: Green chemistry promotes the use of renewable resources, which could lessen dependency on finite fossil fuels and stable supply networks. By emphasizing resource efficiency, companies may promote affordable and clean energy as well as industry, innovation, and infrastructure while becoming more adaptable to changes in the non-renewable resource market.
- Improved Competitiveness for the Environment: Businesses who follow this green chemistry strategy become more competitive by meeting the increased demand from customers for eco-friendly products. This departure from the conventional method enhances branding and contributes to the global sustainable trend of growing market share.

2.3.2.1 Challenges in Green Chemistry from an Economic Perspective

- Initial Investment Costs: Making the switch to green chemistry frequently necessitates large upfront expenditures for infrastructure improvements, new technologies, and research and development [29]. Small- and medium-sized businesses (SMEs) may find this to be a hurdle, making widespread implementation challenging.

- Regulatory Obstacles: The use of green chemical practices may not be sufficiently supported or encouraged by current rules. Businesses may find it difficult to navigate complicated regulatory frameworks that do not place a high priority on sustainability, which could impede innovation. Acceptance in the market despite the increased awareness of sustainability challenges, industries and consumers used to more traditional chemical processes may be resistant. To show the economic potential of green chemistry, this inertia must be addressed through activism and education.
- Limitations of the Supply Chain: The supply chains for sustainable raw materials are frequently somewhat unpredictable, which makes it difficult for businesses to follow the principles of green chemistry. A vital precondition for preserving manufacturing efficiency and meeting sustainability goals is the steady availability of renewable resources. Although it presents a number of difficulties, integrating green chemistry within the SDG framework can be highly beneficial and profitable. Although the biggest obstacle to this goal may be its capacity to overcome obstacles ranging from initial investment to regulatory burdens, market acceptance, and supply chain factors, such integration will lower costs through innovation and create jobs by increasing competitiveness in the efficient use of natural resources.

2.3.3 Societal Reforms Mandating SDGs

Green chemistry plays a significant role in achieving the UN's SDGs by encouraging ecologically friendly methods of producing and using chemicals. Green chemistry must be included in societal reforms to solve the problems of pollution, resource depletion, and public health. Many of the SDGs are strongly related to green chemistry, especially those that emphasize innovation, environmental protection, and sustainable industrialization [30]. The following are important ways that green chemistry supports the SDGs:

- Pollution Prevention: Green chemistry focuses on reducing waste and getting rid of harmful byproducts, which makes the air, water, and soil cleaner. Therefore, by promoting sustainable methods in the production of chemicals, it directly promotes the SDG "Responsible Consumption and Production."

- Utilization of Renewable Resources: Green chemistry supports a low-carbon bioeconomy by enabling a transition from fossil fuel-based resources to renewable ones, such biomass. By supporting sustainable energy alternatives, it helps with both "Climate Action" and "Affordable and Clean Energy."
- Degradable materials: The cycle economy is supported by the design of more straightforward recyclable or biodegradable materials. Because it reduces pollution from persistent items like plastics, this also fits under "life below water" and "life on land."
- Economic Growth and Poverty Alleviation: Green chemistry innovations have the potential to lower production process costs, which will promote economic growth and reduce poverty by creating long-term jobs.
- A number of reforms can be put into place to successfully incorporate green chemistry into social frameworks: Regulations that encourage industries to utilize green chemistry can be put in place by the government. For example, tax incentives could be provided to businesses to promote the use of sustainable materials or procedures.
- Education and Training: Future chemists will be better prepared to develop sustainably if educational programs are improved to include green chemistry principles. Developing a workforce proficient in sustainable practices promotes "Quality Education."
- International Cooperation: Promoting collaborations between governmental, academic, and commercial entities can facilitate the exchange of knowledge and the distribution of resources for green chemistry projects. This supports cooperative efforts for sustainable development, which is consistent with "Partnerships for the Goals."
- Research Funding: More money spent on studies aimed at creating new green technology will stimulate innovation. The achievement of several SDGs pertaining to infrastructure and industry advancements depends on this. Achieving the SDGs requires incorporating green chemistry with social reforms. In addition to addressing environmental issues, green chemistry fosters social justice and economic

development by emphasizing pollution avoidance, renewable resources, and sustainable business practices. To achieve these objectives, it is crucial to put supportive policies into place, improve education, encourage collaboration, and increase financing for research.

2.3.4 SDGs in Perspective—Lessons from Previous Executors

Over the years, several programs and practices have influenced how green chemistry is included in the SDG framework. Analyzing these experiences yields insightful information on successful tactics to be used in the future. Multidisciplinary cooperation, interdisciplinary techniques incorporating chemistry, environmental science, and policy-making collaboration are responsible for the majority of successful green chemistry initiatives. These strategies will provide creative answers to many interrelated SDGs, such as lowering pollution while advancing sustainable business practices [31].

- Put Education and Awareness First: The advancement of green chemistry is significantly influenced by education. A new generation of chemists that are better equipped to tackle environmental issues is produced by programs that incorporate sustainability themes into chemistry courses. This falls under the category of "Quality Education," emphasizing the necessity of reforming education to include sustainability concepts. Appropriate incentives and policy support: It is crucial to have efficient policies that support the adoption of green chemistry practices. Grants and tax breaks can be used to further fund research and development in this area.
- Sustainable Resource Management: Achieving "Responsible Consumption and Production" will need the use of commodities that are sourced sustainably. To lessen the environmental impact, initiatives created in the past could promote renewable resources and offer alternatives to the usage of dangerous chemicals.
- Innovation in Chemical Processes: Successful green chemistry applications have been characterized by the adoption of innovative chemical processes that lower waste and energy consumption. These solutions help both economic efficiency and environmental sustainability by promoting "Decent Work and Economic Growth."

Future initiatives should concentrate on the following to build on these lessons:

- Strengthening International Collaborations: Increasing global cooperation between governments, universities, and businesses can aid in the distribution of resources and the exchange of knowledge for green chemistry projects [32]. "Partnerships for the Goals" is in line with this.
- Creating All-Inclusive Frameworks: Stakeholders can better grasp how to proceed in the realm of sustainable practice with the aid of clear frameworks that outline objectives and guiding principles for green chemistry. The 10 goals outlined by UNEP's Green and Sustainable Chemistry initiative reflect this.
- Encourage Public Awareness: By educating communities about the advantages of green chemistry, you may foster public support for this environmentally friendly approach. Raising awareness can concentrate on how its application promotes environmental, health, and safety preservation. Stakeholders could promote green chemistry projects that are intended to cooperate in the pursuit of the SDGs by applying all of these tactics and learning from prior experiences.

AI disclosure statement: The author confirms that no artificial intelligence (AI)-generated content or AI-assisted writing tools were used in the preparation of this chapter (including summaries and the abstract). All text, analysis, and conclusions are the result of the author's original work.

References

1. Trost, B.M., The atom economy—a search for synthetic efficiency. *Science*, 254, 5037, 1471–7, 1991.
2. Sheldon, R.A., Organic synthesis-past, present and future. *Chem. Ind. (London)*, 23, 903–6, 1992.
3. Sheldon, R.A., Metrics of Green Chemistry and Sustainability: Past, Present, and Future. *ACS Sustain. Chem. Eng.*, 6, 1, 32–48, 2018.
4. Anastas, P.T. and Williamson, T.C., *Green chemistry: designing chemistry for the environment*, ACS Publications, 1996.
5. Anastas, P. and Eghbali, N., Green chemistry: principles and practice. *Chem. Soc. Rev.*, 39, 1, 301–12, 2010.

6. Stories PPS, Office of Pollution Prevention and Toxics, EPA/742/96/002, EPA.
7. Anastas, P.T. and Warner, J.C., *Green chemistry: theory and practice*, Oxford University Press, Green chemistry book, 2000.
8. Constable, D.J., Curzons, A.D., Cunningham, V.L., Metrics to 'green'chemistry—which are the best? *Green Chem.*, 4, 6, 521–7, 2002.
9. Clark, J.H., Green chemistry: challenges and opportunities. *Green Chem.*, 1, 1, 1–8, 1999.
10. Curzons, A.D., Constable, D.J., Mortimer, D.N., Cunningham, V.L., So you think your process is green, how do you know?—Using principles of sustainability to determine what is green–a corporate perspective. *Green Chem.*, 3, 1, 1–6, 2001.
11. Andraos, J. and Sayed, M., On the use of" green" metrics in the undergraduate organic chemistry lecture and lab to assess the mass efficiency of organic reactions. *J. Chem. Educ.*, 84, 6, 1004, 2007.
12. Dicks, A.P. and Hent, A., The E Factor and Process Mass Intensity, in: *Green Chemistry Metrics: A Guide to Determining and Evaluating Process Greenness*, A.P. Dicks and A. Hent (Eds.), pp. 45–67, Springer International Publishing, Cham, 2015.
13. Monteith, E.R., Mampuys, P., Summerton, L., Clark, J.H., Maes, B.U., McElroy, C.R., Why we might be misusing process mass intensity (PMI) and a methodology to apply it effectively as a discovery level metric. *Green Chem.*, 22, 1, 123–35, 2020.
14. Marcinkowska, M., Rasała, D., Puchała, A., Gałuszka, A., A comparison of green chemistry metrics for two methods of bromination and nitration of bis-pyrazolo [3, 4-b; 4′, 3′-e] pyridines. 2011.
15. Kekessie, I., Wegner, K., Martinez, I., Kopach, M.E., White, T.D., Tom, J.K., *et al.*, Process Mass Intensity (PMI): A Holistic Analysis of Current Peptide Manufacturing Processes Informs Sustainability in Peptide Synthesis. *J. Org. Chem.*, 89, 7, 4261–82, 2024.
16. Sheldon, R.A., The E factor: fifteen years on. *Green Chem.*, 9, 12, 1273–83, 2007.
17. Summerton, L. and Constandinou, A., Beyond Mass-based Metrics: Evaluating the Greenness of Your Reaction, in: *Green and Sustainable Medicinal Chemistry: Methods, Tools and Strategies for the 21st Century Pharmaceutical Industry*, L. Summerton, S. HF, J. LC, C. JH (Eds.), The Royal Society of Chemistry, 2016.
18. Shishov, A.Y. and Mokhodoeva, O., Green Chemistry Metrics in Analytical Chemistry. *J. Anal. Chem.*, 79, 5, 487–99, 2024.
19. Zanotti, G., Palmeri, F., Raglione, V., Phthalocyanines Synthesis: A State-of-The-Art Review of Sustainable Approaches Through Green Chemistry Metrics. *Chem.-Eur. J.*, 30, e202400908, 2024.

20. Buonocore, J.J., Choma, E., Villavicencio, A.H., Spengler, J.D., Koehler, D.A., Evans, J.S., *et al.*, Metrics for the sustainable development goals: renewable energy and transportation. *Palgrave Commun.*, 5, 1, 136, 2019.
21. Ribeiro, A.P.C., Martins, LMDRS, Bastos, D.E.N., Cristino, A.F., Galhano dos Santos, R., Chapter 2 - The importance of green chemistry metrics, in: *Handbook of Greener Synthesis of Nanomaterials and Compounds*, B. Kharisov and O. Kharissova (Eds.), pp. 37–62, Elsevier, 2021.
22. Tobiszewski, M., Marć, M., Gałuszka, A., Namieśnik, J., Green Chemistry Metrics with Special Reference to Green Analytical Chemistry. *Molecules*, 20, 6, 10928–46, 2015.
23. Horner, P., Edelman award: 'ORION' delivers success for UPS: revolutionary routing system boosts driver efficiency, cost savings, customer service and the environment. *OR/MS Today*, 43, 22+, 2016.
24. Rana, M.B. and Ahmed Tajuddin, S., Chapter 11: Circular economy and sustainability capability: the case of HM, in: *Upgrading the Global Garment Industry*, Edward Elgar Publishing, Cheltenham, UK, 2021.
25. Brundtland, G.H., Our common future—Call for action. *Environ. Conserv.*, 14, 4, 291–4, 1987.
26. Mitarlis, U.A. and Yonata, B., *J. Technol. Sci. Educ.*, 2022.
27. Shi, L., Han, L., Yang, F., Gao, L., The Evolution of Sustainable Development Theory: Types, Goals, and Research Prospects. *Sustainability*, 11, 24, 7158, 2019.
28. Mishra, M., Desul, S., Santos, C.A.G., Mishra, S.K., Kamal, A.H.M., Goswami, S., *et al.*, A bibliometric analysis of sustainable development goals (SDGs): a review of progress, challenges, and opportunities. *Environ. Dev. Sustainability*, 26, 5, 11101–43, 2024.
29. Chen, T.-L., Kim, H., Pan, S.-Y., Tseng, P.-C., Lin, Y.-P., Chiang, P.-C., Implementation of green chemistry principles in circular economy system towards sustainable development goals: Challenges and perspectives. *Sci. Total Environ.*, 716, 136998, 2020.
30. Bruns, B., Macdonald, I.H., Schneider, B.R., The politics of quality reforms and the challenges for SDGs in education. *World Dev.*, 118, 27–38, 2019.
31. Cheshmehzangi, A. and Dawodu, A., Case Study Reviews: People, Perspective and Planning, in: *Sustainable Urban Development in the Age of Climate Change: People: The Cure or Curse*, A. Cheshmehzangi and A. Dawodu (Eds.), pp. 69–131, Springer Singapore, Singapore, 2019.
32. Younas, S., Shoukat, S., Awan, A., Arslan, S.M., Comparing Effects of Green Innovation and Renewable Energy on Green Economy: The Metrics of Green Economy as Nucleus of SDGs. *Pak. J. Humanit. Soc. Sci.*, 11, 2, 1035–51, 2023.

3

Green Solvents: Fundamentals and Applications

Sinchan Hait, Arshitha Mathew, Anushka Sharma, Akshita Jain and Upasana Mahanta*

Department of Chemical Engineering, Birla Institute of Technology and Science - Pilani, K.K. Birla Goa Campus, Zuarinagar, Goa, India

Abstract

Green solvents have emerged as a favorable alternative in both academia and industry because of the growing demand of solvents on sustainable energy and environment. Considering the demerits of conventional solvents, e.g., flammability, toxicity, high vapor pressure, and lower chemical, electrochemical, and thermal stability, greener or sustainable solvents have been developed during the past few decades. A significant number of organic solvents are also harmful to human health and the environment. Solvents, namely, ionic liquids, deep eutectic solvents, supercritical fluids, switchable solvents, liquid polymers, and renewable solvents have already marked their footprints in various chemical and pharmaceutical applications. Specifically, ionic liquids and deep eutectic solvents have been explored extensively for absorption, solvent extraction, and electrochemical energy storage. Due to their higher cost and viscosity, ionic liquids are progressively being replaced by deep eutectic solvents in most of the chemical engineering research. A variety of techno-economic and environmental factors should be considered to provide the current scenario of green solvent research and development. The proposed book chapter will focus on various types of green solvents useful for chemical and electrochemical applications and their process of synthesis. It will also discuss how several properties of these solvents can affect their performance and improvements required to overcome the limitations.

Keywords: Green solvents, ionic liquids, deep eutectic solvents, CO_2 capture, energy storage

**Corresponding author*: upasanam@goa.bits-pilani.ac.in

Senthilkumar Rathnasamy, Vivek Rangarajan and Umile Gianfranco Spizzirri (eds.) Green Engineering and Sustainable Technology: Tools and Techniques, (71–102) © 2026 Scrivener Publishing LLC

3.1 Introduction

Solvents are an indispensable factor when we discuss green chemistry because of their large quantity used in any industrial manufacturing processes. Most of the presently used industrial solvents are volatile organic compounds, e.g., toluene. Their use leads to high volume of waste, air and water pollution, and also health issues. Having a safer and greener alternative or the use of solvent itself can be the best possible solution to minimize the health and environmental impacts. In any chemical industry, the minor loss of solvents also contributes to organic pollution and solvent removal is, indeed, an energy intensive process. To address the issues with conventional solvents, variety of more environment friendly solvents were designed and developed in recent years. Green chemistry aims to reduce the use of conventional fossil fuel–based solvents and replace them by less toxic alternatives. Green solvents are defined as the reusable, less toxic solvents that can reduce the environmental impact of a chemical reaction or product or process in where it takes part. They were notably studied for solvent extraction, absorption, catalysis, drug delivery, biochemical engineering, and electrochemical and thermal energy storage. Their unique properties (Table 3.1), namely, low vapor pressure, low toxicity, non-flammability, and better thermal and chemical stability, facilitate their engineering applications. Green solvents focus much on their environmental impact and sustainability. However, it is equally important to consider the green and sustainability credentials of the process of solvent synthesis and raw materials. The book chapter includes multiple aspects of some emerging green solvents primarily ionic liquids (ILs), deep eutectic solvents (DESs) and terpene-based solvents.

Table 3.1 Favorable properties of green solvents.

Property	Advantage
Less volatile	It helps to reduce volatile organic compound emissions and air pollution.
Biodegradable	It degrades naturally, which reduces waste management problems.
Derived from renewable sources	It reduces dependency on fossil fuel.
Reusable or recyclable	It increases lifespan and reduces overall consumption of solvents.

ILs are purely ionic compounds with two distinct ions: a) cation that is mostly organic and b) anion that can be either organic or inorganic. The first IL, i.e., ethylammonium nitrate, was discovered in 1914 by Paul Walden [1]. In past three decades, research on ILs has achieved significant milestone in the domain of chemical technology. More than 50,000 articles were published starting from 1996. Their extraordinary thermal stability (>300°C) makes them unique for high-temperature applications. DESs, being a recent addition to green solvents, have emerged as potential solvents for various applications in chemistry and chemical technology. They were notably studied for solvent extraction, absorption, catalysis, drug delivery, and electrochemical and thermal energy storage [2]. They are cost effective than ILs with better electrochemical stability compared to aqueous electrolytes. They are also free from the demerits of high volatility and flammability unlike organic solvents.

One class of secondary metabolites, terpenes and terpenoids, originates from isoprene. Whereas terpenoids contain oxidized functional groups, terpenes are just hydrocarbons. Because these substances are created organically, they can be found all across the natural world. Terpenes are primarily synthesized by plants, and they are essential to several fundamental intra- and intercellular functions. Terpenes are mostly essential oils and find application in cosmetics, food and beverage, and also healthcare products. Terpenes or terpenoids can also be used to synthesize DESs, called natural DES. More discussion on these green solvents can be found in subsequent sections.

3.2 Ionic Liquids (ILs)

The chemical industry has contributed significantly to environmental contamination in the recent decades. Increasing awareness of these issues has led researchers and scientists to broaden more environmentally friendly, much less dangerous products. This category is in line with the principles of green chemistry, which focuses on materials and processes for manufacturing and methods for reducing hazardous waste generated by the use of conventional hazardous chemicals. One promising area is the field of ILs because of their high activity and selectivity. ILs are unique chemical compounds made up with ions and feature very low melting points, typically below 100°C, in contrast to traditional ionic compounds which take the shape of stable crystals at room temperature. Some of the common cations of ILs are imidazolium, ammonium, phosphonium, thiazolium, pyridinium, and triazolium, whereas the anions can range from simple halides to complex tetrafluoroborate or hexafluorophosphate (Figure 3.1). The potential to personalize both the cation and anion makes them flexible to be used for various applications [1]. As already mentioned, the first IL, ethyl ammonium nitrate, was discovered by Paul Walden in 1914. This discovery laid the groundwork for the development of ILs, although it did not grow popular till the 1980s. In the early 1990s, advanced moisture-stable ILs were discovered replacing aluminum chloride with anions like tetrafluoroborate or hexafluorophosphate. This innovation appreciably broadened the scope of ILs, leading to their rapid growth.

ILs are seen as an alternative to traditional solvents used in industrial processes, which are often unsafe due to toxicity, flammability, and high volatility. Because most ILs have melting points lower than room temperature, they are also referred to as room-temperature ILs (RTILs). They have a low vapor pressure and are not easily evaporated and thus they are perceived to be safer and also reduce environmental contamination [3]. These properties of ILs make them useful in a wide range of applications. It has been used in industry to enhance catalyst recovery, improve separation, and increase the efficiency of various chemical reactions. However, due to the high internal friction of fluids, and ILs have a higher viscosity than most molecular solvents, which limits their industrial uses.

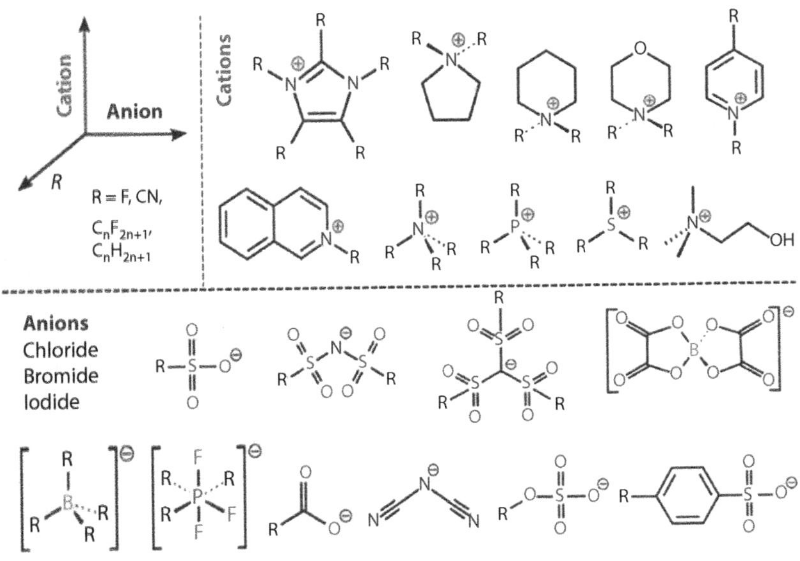

Figure 3.1 Some commonly reported anions and cations. R = functional group. Reprinted with permission [4].

3.2.1 Properties of Ionic Liquids

3.2.1.1 Low Melting Point

The large size and asymmetry of the constituent ions imparts significantly low melting points to ILs. This helps them to be loosely packed without forming solid crystals at room temperature. Hence, for low-temperature applications, ILs can replace other solvents, which might undergo solidification. The presence of water and other solvent may further decrease the melting point of ILs. However, it increases with increase in alkyl chain length.

3.2.1.2 High Thermal Stability

Due to the excellent thermal stability (>300°C) of the available ILs, they can be easily used for high-temperature applications without any thermal degradation.

3.2.1.3 Low Vapor Pressure

ILs evaporate slowly and create insignificant vapor to cause any hazard. Because of this feature, ILs commonly do not release any pollutant to the environment. Further, loss of solvent during operation is also minimal.

3.2.1.4 Non-Flammability

ILs benefit industrial applications with a lower chance of a fire explosion due to their non-flammable nature.

3.2.1.5 Solubility and Miscibility

ILs can attract metals and organic compounds and form a solution. Moreover, ILs can be designed for dissolving other compounds as well and used as solvents for wide operations.

3.2.1.6 Electrical Conductivity

ILs are highly conductive due to their structural configuration, which is composed of ions. This property makes them ideal for various electrochemical applications.

3.2.1.7 *Chemical Stability*

ILs are not likely to react with other substances, which ensures that they remain unchanged in a variety of conditions.

3.2.2 Applications of Ionic Liquids

3.2.2.1 *Electrochemical Applications*

The two more appealing characteristics of ILs or RTILs for their employment in electrochemical storage devices are their wide operating potential window and reasonable ionic conductivity. They are also investigated as a media for electrosynthesis of polymer. Further, ILs find application in electrochromic devices, electrochemical sensors, and electrochemical biosensors. Majority of ILs used in electrochemical devices belong to imidazolium

Table 3.2 Summary of ILs' properties and their role in electrochemistry.

Property	Role
High ionic conductivity	It allows ions to pass through the electrolyte ions more efficiently and is important for electrochemical reactions and ion diffusion.
Wider electrochemical window	A wide range of potentials can be applied without decomposing the electrolyte. The more the potential window of the electrolyte, the higher the energy density of the supercapacitor or battery.
Low vapor pressure	It reduces solvent loss, reduces environmental impact, and improves product efficiency.
Lewis acidity (for chloroaluminate RTILs)	The adjustable Lewis acidity allows the dissolution and stabilization of compounds that are often difficult to handle in conventional formulations, facilitating the study and application of these compounds.
Electrodeposition capability	It facilitates the fabrication of thin layers of various metals on the electrode surface, enabling research and device development of new materials.
Thermal stability	The excellent thermal stability of majority of ILs facilitates high-temperature applications of electrochemical devices or processes.

category [5]. Table 3.2 explains how several properties of ILs induce their electrochemical applications. Nevertheless, the primary drawback of pure ILs is the higher viscosity at ambient temperature than organic electrolytes, which imparts more resistance to ion movement. Despite the demerits, previously reported research shows that supercapacitors with ILs as electrolytes results in increase in energy density when compared to aqueous electrolytes [6–9]. Addition of organic solvents or supporting electrolytes helps to reduce the viscosity of ILs and hence enhance ion mobility. ILs are also considered as suitable candidates to synthesis gel polymer electrolytes for energy storage devices [10, 11]. Molecular dynamic (MD) simulation studies on IL clears the ion transport mechanism at molecular level [12]. Electrodeposition and electrocatalysis are also two major electrochemical applications of ILs [13]. It is possible to electro deposit most of the less reactive metals in ILs that were previously electrodeposited in aqueous solutions. One benefit of using ILs rather than aqueous solution is that it is simpler to electrodeposit metals utilizing direct electrodeposition without the need for further annealing due to high-temperature stability of ILs.

3.2.2.2 Gas Absorption

ILs are well acknowledged as innovative solvents for gas absorption and separation applications for the separation of basic and acid gasses in industrial processes. ILs can potentially resolve the issues brought on by

Table 3.3 Properties affecting gas absorption in ILs.

Property	**Advantage**
Insignificant vapor pressure	Low volatility lessens the impact on the environment and loss
Increased chemical and thermal stability	Adoptable at high temperatures and impervious to the gas' chemical reaction
Adjustable structure	Changing the cations and anions makes building it for a gas absorption process possible
Minimal melting point	Many are liquid at normal temperature
Elevated polarity	Excellent polar compound solvent
Solvation property	Able to dissolve a broad variety of both organic and inorganic substances
Not flammable	Safer to work with than organic solvents

the conventional organic solvent techniques, such as their high energy need and pollution. Gases, namely, CO_2, SO_2, H_2S, NH_3, N_2, and CH_4, are commonly absorbed and can be separated by conventional and functionalized ILs [14]. Computational tools, e.g., molecular dynamic simulation [15, 16], quantum chemical calculations [17], and conductor like screening model (COSMO) for real solvents, were [18, 19] also used to study the molecular mechanism and thermodynamics of gas solubility in ILs. These studies explained the insights on gas absorption rate based on their molecular structure, polarity, and solvation capacity (Henry's constant). Imidazolium-based ILs were also reported for separation of biogases [20]. The physio-chemical properties that influence gas solubility are listed in Table 3.3. However, high viscosity and cost of ILs cannot be ignored while considered for industrial applications.

3.2.2.3 Drug Delivery

One of the emerging applications of the ILs is developing a new drug delivery system. ILs improve drug formulation by enhancing its solubility, stability, drug release, increased drug loading capacity, and permeability [20–24]. Even the thermodynamic properties of the IL play a crucial part in this regard. In this aspect, COSMO-RS better forecasts drug solubility, activity coefficient, and equilibrium data (phase) [22]. Some surface-activated IL is good choice for drug delivery because of its membrane-like structure, which encapsulates the drug molecules regardless of their polarity and release depending upon the pH of the surrounding media [25]. Some important parameters affecting the drug delivery process are discussed in the table below (Table 3.4).

3.2.2.4 Catalysis

In 1986, for the first time, IL was reported as a catalyst in Friedel–Craft acylation reaction [26]. Later, their catalytic activity was studied extensively in the past three decades. Because of their low volatility, ILs can be recovered easily after the completion of the reaction. ILs can be promising catalysts for synthesis of metal nanoparticles, transition metal catalysis, alkene hydrogenation, oxidation, hydroformylation, and olefin-dimerization [27]. Further, organocatalysis also explored the potential of ILs as a catalyst with other organic compounds [27]. The organocatalysts are involved in Diels–Alder cycloaddition reactions through hydrogen bonding formation. Condensation reactions were also carried in chloroaluminate (III) ILs.

Table 3.4 Properties influencing drug delivery.

Property	Overview	Significance
Biocompatibility	Biological systems can safely interact with IL without any adverse side effects.	Toxic effects and adverse immunological responses should be prohibited
Improved solubility	Capable of dissolving hydrophilic and hydrophobic compounds.	Overcome the problem of low water solubility of the composition of medication
Adjustability	Tunable structural modification changes physio-chemical properties, e.g., polarity, hydrophobicity, hydrophilicity, and viscosity.	It allows the tuning of the drug delivery systems to lead to various therapeutic choices
Stability	Strong ionic bonds increase a material's resistance to chemical and thermal degradation.	Increases the sterilization potential and long-term stability of the drug during its delivery, storage, and transit
Minimal vapor pressure	Low tendency for evaporation.	Less hazardous to formulate and handle while delivering
Personalized delivery	It can be programmed to release in response to temperature, pH, and enzymes.	Increases activity and reduces the systemic effects by delivering the drug specifically to the target sites, thus reducing the local adverse effects
Two-fold purpose	Besides acting as solvent properties, some ILs have intrinsic pharmaceutical properties.	By simplifying formulations, it provides the opportunity for positive synergisms

They were explored as co-catalysts or catalyst support for various reactions. Occasionally, ILs can also act as a solvent for reaction. Table 3.5 lists the properties of ILs with their influence in catalysis.

Table 3.5 Properties of ILs affecting catalysis.

Property	Description	Importance
Solvent properties	Its ability to dissolve many substances, ranging from organic to inorganic materials, has earned it this name.	It offers a homogeneous reaction environment to the catalyst and reactants.
Stability and reusability	Chemically and thermally stable, resistant to decomposition.	This enables repeat use without high-performance loss and minimizes waste and associated costs.
Customizable properties	The alteration in cations and anions can control the physical and chemical properties.	Makes the ILs amenable to optimization for any specific catalytic process.
Low vapor pressure	Minimum tendency to evaporate, reducing loss of volatile components.	Prevents loss of reactants, products, and catalysts.
Selectivity	More specific interaction with reactants and catalysts indicates that it may be possible to influence the pathways of reactions.	The tool allows for the formation of certain products and the suppression of side reactions.
Facilitates multiphase reactions	It acts as the catalysis medium in multiphase systems using a solvent.	Improved mass transfer between phases, enhancing reaction rates.
Environmental benefits	Replace traditional, often hazardous solvents.	It reduces environmental pollution and health risks.

3.3 Deep Eutectic Solvents (DESs)

A DES is primarily a eutectic mixture of hydrogen bond acceptor (HBA) and a hydrogen bond donor (HBD), mixed in a particular mole ratio, and has a low melting point [2]. DESs can also be termed as designer solvents due to the flexibility of assembling desired HBA:HBD combinations. The review by Tome *et al.* explains the dominance of choline chloride (ChCl)–based DESs' research starting from the 21st century [28]. The authors reported several applications of DES including polymer science, metals, and materials. Previously, a comparative study on the physicochemical properties of various DESs based on ChCl, ammonium, and phosphonium salts was reported by Gracia *et al.* The authors also analyzed use of DESs for gas separation mainly absorption of CO_2 and SO_2 [29]. Malaquias *et al.* used ChCl:urea to perform electrodeposition of Cu-In alloy for photovoltaic applications [30]. ChCl with urea and glycerol as HBDs was also employed as surfactants for enhanced oil recovery [31]. Multiple DESs containing $ZnCl_2$ as HBA were characterized for electrochemical applications by Bagh *et al.* [32]. Another DES with ChCl as HBA and p-chlorophenol as HBD at 1:2 molar ratio was used by Farajzadeh *et al.* for extraction of solvent for polycyclic aromatic hydrocarbons from water [33]. However, the constraint of increased viscosity of most conventional DESs limits their applications. To overcome this constraint, researchers started working on low viscous DESs as alternatives to conventional solvents. In this work, the authors will focus on DESs used in the domain of CO_2 capture and electrical double layer capacitor.

3.3.1 Types of Deep Eutectic Solvents

DESs are primarily composed of an ionic compound which acts as HBA (Cat^+X^-) with different HBDs (Figure 3.2). Table 3.6 reflects different types of conventional DES according to the composition. A new class of DES (DES-V) composed of organic molecules has been recently studied for heat storage applications [34, 35].

3.3.1.1 *Type I: (Salt + Metal Compound)*

Type I DESs are liquids over a range of temperatures and have strong metal-binding properties. Precious metals can be extracted from electronic waste and can also be used to capture CO_2 using these solvents.

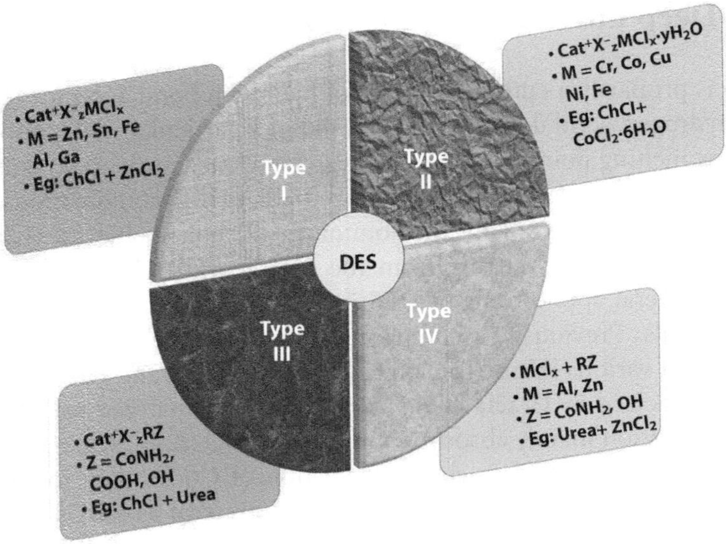

Figure 3.2 Primary categories of deep eutectic solvents.

Table 3.6 Different types of DES.

Type	General formula	Terms	Significance of z
I	$Cat^+X^-zMCl_x$	M = Zn, Sn, Fe, Al, Ga, In	Z number of metal chloride
II	$Cat^+X^-zMCl_x.yH_2O$	M = Cr, Co, Cu, Ni, Fe	Metal chloride hydrate
III	Cat^+X^-zRZ	Z = COOH, OH, CONH	Hydrogen bond donor
IV	$MCl_x{}^+RZ$	M = Al, Zn Z = $CONH_2$, OH	Consist of metal chloride and hydrogen bond donor

3.3.1.2 *Type II: (Salt + Metal Compound + Water)*

The water present in this type of DES enhances its ability to form hydrogen compounds. They are used in the production of many products like biodiesel fuel and also used in the wastewater (WW) treatment process.

3.3.1.3 *Type III: (Salt + Organic Molecule)*

Type III DESs are sustainable solvents as they use only organic solvents like sugars and alcohols. They are used in the pharmaceutical and cosmetic industry.

3.3.1.4 *Type IV: (Salt + Organic Molecule + Metal Compound)*

Type IV DESs combine metal compounds with organic molecules combining the advantages of both type III and type I DES. Their unique structure enables them in degrading pollutants or separating valuable substances from mixtures. They offer specialized solutions for a variety of applications.

3.3.1.5 *Type V: (Organic Molecule + Organic Molecule)*

These are the newest types of DESs and are unique because they do not use salts and use only organic molecules. Although they are still in their early stages of development and research is ongoing, their special structure shows potential for use in thermal energy storage.

3.3.2 Properties of Deep Eutectic Solvents

3.3.2.1 *Melting Point*

The very low melting point of DESs in relation to their constituent parts is one of its distinguishing features. This results from the development of an H-bond between HBD and HBA. The ultimate melting point is determined in large part by the choice of HBD, its structure, and the molar ratio with the salt.

3.3.2.2 *Density and Viscosity*

The density of the DESs is higher than that of the water due to stronger intermolecular H-bonds between the HBA and HBD, mainly at room temperature. Unlike the density of the DESs, viscosity is also influenced by the strength of the H-bond formation. In addition, the lower density and viscosity occur due to weaker and/or less H-bond formation.

3.3.2.3 *Ionic Charge*

Ionicity determines the number of the ions available for conductivity. DESs may contain neutral ions or clusters that do not contribute to charge transfer, therefore impacting the overall ionicity. In addition, higher ionicity means higher conductivity, which makes them an ideal for electrochemical application. Tightly packed ions in highly ionic DESs can lead to higher density.

3.3.2.4 *Ionic Conductivity*

It is important to observe that viscosity and ionic conductivity are associated. Low conductivity is seen in very viscous DES, and vice versa. On the other hand, conductivity advantages with rising temperatures. The conductivity of DESs is likewise extensively influenced by means of composition. Higher conductivity results in improved mobility of the ions for a low viscous DES.

For a variety of DES packages, melting point, density, viscosity, and conductivity are critical. Nonetheless, additional traits that include toxicity, biodegradability, and chemical and thermal stability also are vital. To properly comprehend and take advantage of those trends, sizable observation and optimization are required.

3.3.3 Preparation of Deep Eutectic Solvents

The steps involved in the process of DES preparation are discussed as follows. The process is less energy intensive compared to other laboratory solvent synthesis processes. Unlike ILs, DES preparation process does not produce any side product, and, hence, it is cost effective.

a. Grinding: It is usually done using a mortar and pestle to reduce the particle size of the solid components. This enhances physical interaction between HBA-HBD and reduces heating rate.
b. Heating: The second step is to mix HBD and HBA in the desired molar ratio and then the mixture is heated at elevated temperature (70°C–80°C) for 12–24 hours. To obtain a homogeneous eutectic mixture, continuous stirring is essential while heating. The resultant DES is obtained after cooling the homogeneous mixture.
c. Ultrasonication: This can be done as a post treatment after completion of heating to accelerate mixing and hydrogen bonding formation.
d. Microwave Irradiation: This step can be followed instead of simple heating. It reduces preparation time.

The additional factors influencing DES preparation:

- Purity: The final properties of DES can be affected by impurities in HBA or HBD. Hence, purification of raw materials is very important.
- Moisture Content: The presence of water can influence melting point and stability, which requires drying of moist HBA and HBD.
- Heating Time: Overheating of the mixture may sometimes decompose the samples or induce side reactions.

3.3.4 Applications of Deep Eutectic Solvents

3.3.4.1 Deep Eutectic Solvents for CO_2 Capture

Anthropogenic CO_2 emissions are the major contributor to the greenhouse gases (GHGs). In 1986, Svante Arrhenius provided the initial evidence of a human-caused greenhouse effect. He initiated the analysis of the relation between the rise in atmospheric CO_2 levels and Earth's surface temperature. Arrhenius proposed that, if the concentration of CO_2 doubled, then it would result in a temperature increase of 4°C–6°C [36]. The increase in GHG is unfavorable for climate change. Depending upon the source of CO_2 capture, they were identified as pre-combustion, post-combustion, and oxyfuel combustion. The fuel is vaporized to produce combustible gas, which is succeeded by CO_2 removal in pre-combustion. This process produces a high concentration of CO_2. Due to this, physical absorption becomes easier, and, thus, solvent regeneration becomes simple, which is explained later in this chapter. In pre-combustion, the fuel is transformed into a combustible gas before burning. In contrast, in post-combustion, the power plant is equipped with CO_2 collection technology following fossil fuel burning. However, post-combustion produces less CO_2 than pre-combustion, increasing its energy penalty and the cost. In contrast to pre- and post-combustion, oxyfueling entails burning the fuel in an oxygen-rich gas flow while primarily creating CO_2, which is helpful for easy CO_2 extraction. Besides that, it produces a higher amount of CO_2 and eliminates NOx as it uses a pure oxygen atmosphere. However, CO_2 emission from oxy-fuel combustion plants is far less than its previous counterparts [37].

Several technologies have been developed to address this, such as absorption (chemical and physical), adsorption (chemical and physical), membrane, gas-hydrate crystallization, cryogenic separation, and direct air capture [37]. Absorption gives a good yield, but the solvents' volatility, toxicity, and regeneration remain the biggest challenge. Meanwhile, using membranes overcomes the regeneration process and provides a simple operational setup. However, fouling and hindered mass transfer (MT) make it unsuitable. The cryogenic separation process is highly energy-intensive and applicable only to high carbon concentrations (>90%). This limits its application at a larger scale. Moreover, adsorption gives a low yield compared to absorption, but regeneration of the components is easy. Nevertheless, low-energy demand, less viscous (chemical), and cheap availability make it an ideal solution for the CO_2 capture process.

Predominantly, solvents such as organic and ILs have been used for carbon capture. However, these solvents adversely affect the environment and human health. Therefore, a greener and more sustainable approach is needed to minimize this implication. The high vapor pressure and toxicity of the organic solvents make them highly volatile, thus causing them to lose solvents during the process and making them unsafe. The alkanol amines regenerated at 100°C–150°C, which requires high energy demand and, hence, a surge in cost [38]. Nonetheless, amines undergo thermal degradation at this high temperature, which releases harmful chemical components and needs a separate water wash setup. On the other hand, ILs, which are molten salts of cations and anions, overcome the volatility, flammability, and thermal stability. The benefit of the ILs is that they have negligible vapor pressure and regeneration of solvent easy. Even though they have been successfully established for CO_2 capture, the cost of these solvents becomes con. Moreover, CO_2 loading capacity is influenced by the solvent characteristics, CO_2 partial pressure, and temperature.

Recently, DES, which is a mixer of HBD and HBA, emerged as an alternative green solvent. The DES's ease of design makes it a perfect candidate for CO_2 capture. Li *et al.* first introduced that DES can be used for CO_2 capture [39]. Since then, DES for CO_2 capture has proliferated. Physicochemical properties, such as vapor pressure, viscosity, density, refractive, hydrophobicity, hydrophilicity, and temperature profile, can be tailor-made depending on the precursors used (HBD and HBA). These properties directly influence the MT process during CO_2 absorption and can be eliminated by using DES. For instance, organic solvents such as amines are highly viscous and will work as a rate-limiting factor in MT. To improve the gas solubility, the solvent must be less dense so that CO_2 molecules have more space to interact. It is well-known that viscosity and density are directly proportional to temperature, so the operating temperature needs to be raised for a highly viscous solvent to decrease viscosity and increase CO_2 absorption. Not only that, pH also plays a crucial role in this process [38]. As CO_2 is acidic, it needs a basic solvent to improve its solubility. This can be attributed to the fact that CO_2 is highly soluble in amine, mainly due to its basic nature. However, as mentioned earlier, the degradation of amines produces harmful pollutants.

3.3.4.2 CO_2 Capture Mechanism

Solvent absorption of CO_2 occurs in two ways: a) chemically, i.e., chemisorption, and b) physically, i.e., physisorption. Detailed description of the processes can be found as follows.

3.3.4.2.1 Chemisorption

Chemisorption gives higher solubility *via* carbamate formation. The following reactions show how chemisorption occurs in amines through the zwitterion formation:

$$R_1R_2NH + CO_2 \rightarrow R_1R_2NH^+COO^-$$

$$R_1R_2NH^+COO^- + B \rightarrow R_1R_2NCOO^- + BH^+$$

B represents any base and contributes to the deprotonation during the zwitterion process. This scheme is well established for CO_2 capture. CO_2

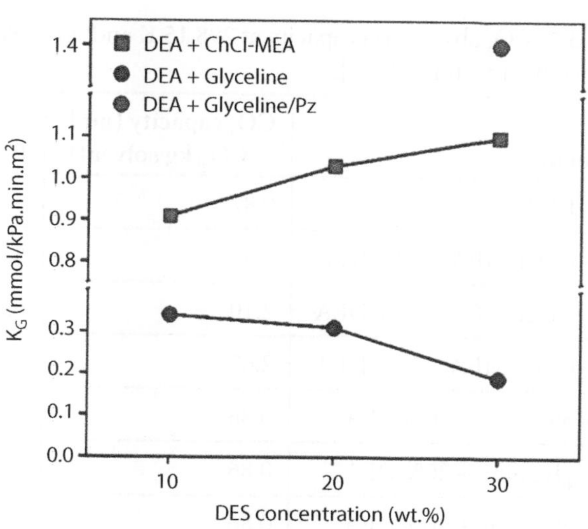

Figure 3.3 Influence of MT co-efficient at 20% DES concentration. Reprinted with permission [40].

reacts with the base in a 2:1 stoichiometry and forms a carbamate complex. For hydrophilic components (e.g., amines), the presence of water improves the gas solubility by reducing the solvent's viscosity, increasing free volume, and better interactions among CO_2 molecules and the solvent components. DESs also follow the similar mechanism of CO_2 capture. A study by Karami *et al.* used choline chloride (ChCl) as HBA and monoethanolamine (MEA), glycerol (Gly), and piperazine (Pz) as HBD from DES [40]. DESs [ChCl-MEA (1:7), ChCl-Gly (glyceline; 1:2), and glyceline/Pz (1:2:1)] were mixed with diethanolamine (DEA) in the 10–30 wt% range. In glyceline/Pz DES, they proposed a pathway for carbamate formation, reacting with Pz's NH group. The gas flow regime influences MT, hydrodynamic parameters, and gas solubility in bubble reactors. However, other parameters such as reactor dimension, distributor design, and liquid viscosity also impact the flow regime transition [40]. They designed the reactor to meet a homogeneous flow regime during all operations. Nonetheless, the supercritical gas velocity is characterized by homogeneous small bubble size, reduced rising velocity, and restricted interaction among bubbles, leading to efficient MT. Bubble size distribution (BD) shows micro-bubbles, therefore, high gas-liquid interfacial area and higher residence time. In addition, the partial substitution of water by DES improves conductivity and reduces coalescence-induced bubbles. This mainly occurs as electrolytes are good suppressors of coalescence [40]. On the contrary, although 20% DEA + 30% ChCl + MEA was less conductive compared to 20% DEA + 30%

Table 3.7 CO_2 absorption capacity at 298.15 K and 0.1 MPa; enhancement factor (E) [40].

Sorbent	CO_2 capacity (mol CO_2/kg solvent)	E
20% DEA	0.87	3.03
10% ChCl-MEA + 20% DEA	1.55	5.41
20% ChCl-MEA + 20% DEA	2.10	6.82
30% ChCl-MEA + 20% DEA	2.67	8.30
10% glyceline + 20% DEA	0.86	2.04
20% glyceline + 20% DEA	0.86	2.02
30% glyceline + 20% DEA	0.86	1.50
30% glyceline/Pz + 20% DEA	1.53	13.46

glyceline, it had smaller bubbles, probably due to the availability of reactant molecules. By coalescence, the most significant gas holdup was observed in glyceline-DEA and ChCl-MEA-DEA due to the bubble size. Besides that, ChCl-MEA showed a higher adequate sorption time (when the outlet and inlet CO_2 concentration ratio reached 0.1) than the glyceline-DEA solution (Table 3.7). Still, it was similar to the glyceline/Pz-DEA. Similarly, CO_2 capture capacity was higher in ChCl-MEA (2.67 mol CO_2/kg solvent; 30 wt% ChCL-MEA in DEA and 30 wt% glyceline/Pz-DEA has a capacity of 1.53 mol CO_2/kg solvent [40]. Unlike this, the absorption rate also showed a similar trend; with increasing ChCl-MEA or glyceline/Pz, the CO_2 rate of absorption rose. The chemical reaction between CO_2 and DES shifts toward the gas-liquid interface and, thereby, a steeper concentration distribution curve, reducing MT resistance on the liquid side.

Increasing glyceline concentration to 30 wt% negatively affected the absorption rate. This can be attributed to glyceline enhancing interfacial area (favoring absorption rate), but higher viscosity reduces diffusivity [40]. Moreover, from Figure 3.3, it is evident that the supplementation of DES improves overall MT co-efficient.

As ChCl-MEA reacts with CO_2, it provides a high enhancement factor; it increases from 3 to 8.3 by substituting ChCl-MEA to DEA from 10 to 30 wt%. However, it decreases in glyceline (up to 50%), but a higher value was observed for the 30% glyceline/Pz + 20% DEA system. Even though the ChCl-MEA-DEA system gave high solubility, thermal stability was a significant drawback in regeneration; glyceline/Pz-DEA was much more stable up to 180°C. Also, post-CO_2 absorption, glyceline/Pz-DEA and glyceline-DEA stability were compromised, making regenerating easy [40].

3.3.4.2.2 Physisorption

Even though chemisorption gives a high absorption rate, solvent regeneration becomes challenging and energy-intensive. Therefore, an alternative must be considered. Thus, the current research focuses on the physisorption, i.e., CO_2 molecules with solvent by weak forces. These types of DESs are hydrophobic. A study by Siani G. *et al.* studied the physical absorption of CO_2 in trimethylglycine (TMG), oxalic acid dihydrate, glycolic acid, and phenylacetic acid–based natural DESs (NDES) [38]. The absorption experiment was carried out till 4 MPa at temperature starting from 298.15 K to 333.15 K. TMG/phenylacetic acid NDES gave the highest solubility at all pressures, i.e., 45.5 ± 1.7 mg CO_2/g NDES (at 2.5 MPa, 313.15 K). As expected, with the temperature rise (333.15 K), despite the pressure increasing, solubility reduces to 33.7 ± 1.1 mg CO_2/g NDES [38] (Table 3.8).

Table 3.8 CO_2 absorption in NDES after 1-hour (2-hour) interval from gas induction at different temperatures and pressures [38].

Sorbent	Pressure (MPa)	CO_2 capacity (mg CO_2/g NDES) Temperature (K)		
		298.15	313.15	333.15
TMG/oxalic acid	0.1	0	0.35 ± 0.02	0.32 ± 0.02
dihydrate	1	0.48 ± 0.03	0.45 ± 0.03	0.18 ± 0.03
	4	0.24 ± 0.03	0.52 ± 0.04	1.72 ± 0.11
TMG/glycolic acid	0.1	0.85 ± 0.06	0.54 ± 0.04	0.24 ± 0.03
	1	7.64 ± 0.31	2.91 ± 0.18	3.10 ± 0.15
	4	9.15 ± 0.55	2.84 ± 0.19	1.68 ± 0.11
TMG/phenylacetic acid	0.1	1.15 ± 0.07	1.18 ± 0.07 (15.9 ± 0.7, 2 hour)	1.15 ± 0.07 (12.0 ± 0.4, 2 hour)
	1	9.9 2± 0.43	13.1 ± 0.5 (28.3 ± 0.7, 2 hour)	12.3 ± 0.6 (20.5 ± 0.6, 2 hour)
	2.5	13.6 ± 0.5	21.6 ± 0.9	20.5 ± 0.8
	4	14.5 ± 0.5	38.1 ± 1.2 (45.5 ± 1.7, 2 hour)	32.0 ± 1.1 (33.7 ± 1.1)

In this study, in NDESs, HBA was kept the same, and HBDs varied, indicating HBD's influence on CO_2 absorption. The lower solubility of the other two NDES can be explained by the acid dissociation coefficient's negative log (*pKa*) values of the individual components (HBDs). As the HBDs have a carboxylic functional group, it is expected that they will have lower *pKa* values, e.g., phenylacetic acid (*pKa* = 4.31), oxalic acid (*pKa* = 1.46), and glycolic acid (*pKa* = 3.6). As discussed earlier, the pH of the solvent greatly influences the CO_2 solubility, as the CO_2 shows a high affinity towards the basic solvent. So, a higher *pKa* value means lower acidity; thus, phenylacetic acid gave better solubility than its counterpart. In NDESs, CO_2 is attached to the NDES *via* weak forces such as electrostatic and dispersion interactions [38].

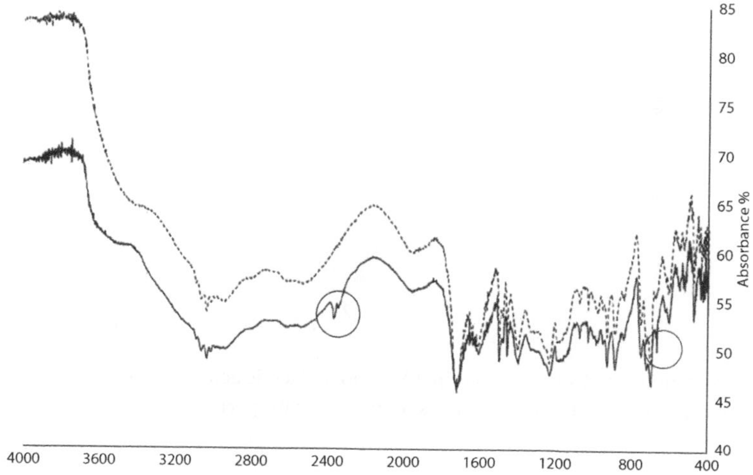

Figure 3.4 FTIR spectra of TMG/phenylacetic acid before (dashed line) and after exposure (full line) with CO_2. Reprinted with permission [38].

Nevertheless, when considering dicarboxylic oxalic and glycolic acids, the additional carbonyl and alcoholic groups enable the creation of a robust, complex intra- and intermolecular H-bond network between the acids, leading to reduced CO_2 physisorption. In addition, water molecules in TMG/oxalic acid DES form the H-bond network and limit the CO_2 binding site in NDES. The third NDES, containing carboxylic acid with an aromatic ring, demonstrated π-π interactions between aromatic and non-aromatic groups. This infers the reason for TMG/phenylacetic acid giving high solubility, and, thus, π delocalized systems are possibly the best candidates for CO_2 physisorption [38]. In addition, there is no chemical interaction between NDES components and CO_2. Still, the gas is physically absorbed, as evidenced by the appearance of the CO_2 asymmetric stretching band (2350 cm^{-1}) and the CO_2 active bending band (668 cm^{-1}) in the spectrum after exposure (Figure 3.4).

The NDES components absorb heavily in the same region, resulting in a faint band at 668 cm^{-1}. The density trend of the NDES is as follows: TMG/oxalic acid>TMG/glycolic acid > TMG/phenylacetic acid. The free volume of DES is inversely proportional to its density; this indicates TMG/phenylacetic acid has a large free volume compared to its counterparts. The higher free volume provides high interaction space, so TMG/phenylacetic acid gave the highest solubility [38]. At lower temperatures, it takes a longer time for the system to reach equilibrium. The NDES was successfully recycled at 343.15 K for three cycles under vacuum (Figure 3.5).

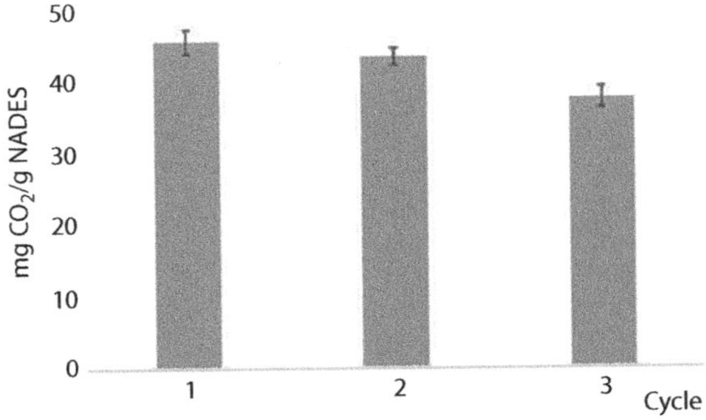

Figure 3.5 Amount of CO_2 absorbed in TMG/phenylacetic acid DES for three consecutive absorption/desorption cycles. Reprinted with permission [38].

3.3.5 Green Credentials

DESs are being well recognized in green chemistry due to their low vapor pressure and biodegradability. However, naming them as "green solvents" may not be justifiable in each case. The green features of DESs primarily depend on their composition and the output of the processes in which they are used. Type III DESs, consisting of urea and glycerol, etc., possess lower toxicity compared to other metal salts containing DESs. Despite the advantages of biodegradability, low vapor pressure, cost-effectiveness, and stability, there are challenges to deal with. In electroplating, the high viscosity of DES consumes more time, leading to higher volume of waste. Moreover, chlorinated byproducts from DESs are the major pollutants to the environment. In this regards, to minimize the waste nanofiltration can be one possible approach to be followed. In order to truly name DESs as "green solvents," it is important to analyze the effects on environment. Even though their demerits, DESs shows potential in multiple fields including catalysis, electrochemistry, absorption, liquid-liquid extraction (LLE), and drug delivery. The ever increasing research on DESs and its applications supports their sustainability. Therefore, while considering DESs as greener solvents, careful assessment and in-depth research is essential.

3.4 Terpenes and Their DES Derivatives

Among several natural secondary metabolites, terpenes are exceptionally diverse and prevalent. In a primary classification, secondary metabolites are divided into three main groups: nitrogen-containing compounds

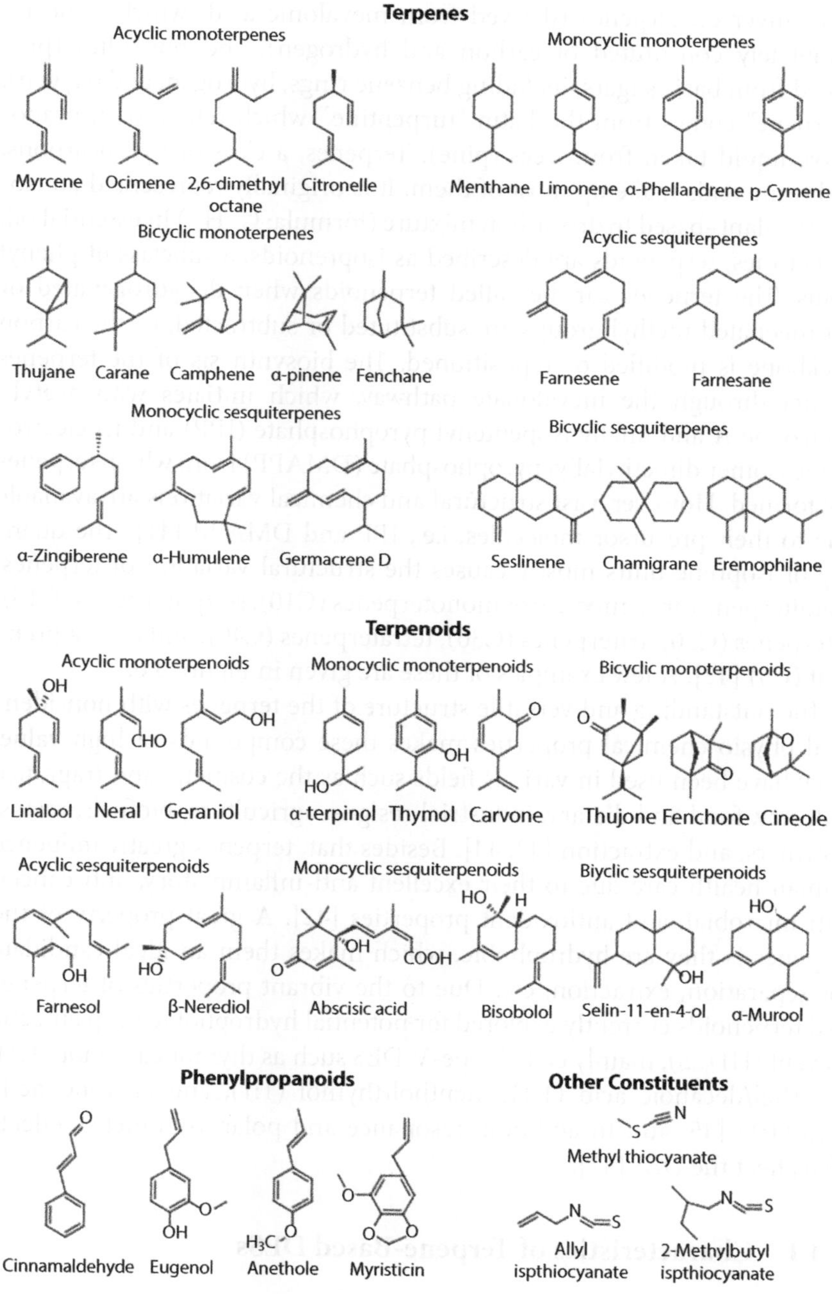

Figure 3.6 Terpene chemical structure. Reprinted with permission [39].

(very diverse), terpenes (derived from mevalonic acid, which is nearly completely constituted of carbon and hydrogen), and phenolics (produced from basic sugars, including benzene rings, hydrogen, and oxygen). "Terpene" comes from the Latin "turpentine," which refers to an evaporative liquid taken from trees (pine). Terpenes, a class of hydrocarbons, and resin acids make up most of them. It is originally represented an isometric plant–based hydrocarbon mixture (formula: $C_{10}H_{16}$) in essential oil. Sometimes, terpenoids are described as isoprenoids, a subclass of phenyl lipids. The terpenes can be called terpenoids when dehydrogenated or hydrogenated methyl groups are substituted or subtracted, or the carbon backbone is modified or repositioned. The biosynthesis of the terpenes occurs through the mevalonate pathway, which initiates with acetyl–coenzyme A and finally isopentenyl pyrophosphate (IPP) and its electrophilic isomer dimethylallyl pyrophosphate (DMAPP) from which terpenes are formed. However, vast structural and chemical variations are available due to their precursor molecules, i.e., IPP and DMAPP [41]. The quantity of isoprene units mostly causes the structural variation of terpenes. Hemiterpenes are a mixture of monoterpenes (C10), sesquiterpenes (C15), diterpenes (C20), triterpenes (C30), tetraterpenes (C40), and one isoprene unit (C5) [42]. A few examples of these are given in Figure 3.6.

The outstanding and versatile structure of the terpenes with non-identical physio-chemical properties makes these compounds of high value. They have been used in various fields such as the cosmetic and fragrance industry, food and dietary, industrial resigns, agriculture, biofuels, resins, polymers, and extraction [43, 44]. Besides that, terpenes greatly influence human health care due to their excellent anti-inflammatory, anti-cancer, anti-microbial, and antioxidant properties [42]. A great property of the terpenes is they are hydrophobic, which makes them an ideal candidate for separation, extraction, etc. Due to the vibrant properties of terpenes and terpenoids currently explored for potential hydrophobic deep eutectic solvent (HDES), mainly called type-V DES such as thymol/camphor (1:1), menthol/decanoic acid (1:1), menthol/thymol (1:1), and carvone/lactic acid (1:1) [45, 46]. In addition, resonance and polar asymmetric effects also affect the DES [47].

3.4.1 Characteristics of Terpene-Based DESs

3.4.1.1 Viscosity

Viscosity is an important parameter dominated by the van der Waals force, and the hydrogen bonds play an essential part in fluids' motion [48].

Ge *et al.* showed that, in thymol/GML DES, as the amount of GML increases, the viscosity of the DES also increases [48]. In another study by Wazeer *et al.*, six different HDES (hydrocinnamic acid/decanoic acid, menthol/thymol, thymol/decanoic acid, menthol/decanoic acid, thymol/coumarin, and thymol/camphor) for removing creosol from WW [46]. All the studied HDES had viscosity <40 mPa·s at room temperature. The lower viscosity is ideal for industrial processes and LLE. Viscous liquids are a challenging task for industrial processes. Al-Bodour *et al.* noted that a higher ratio of HBD increased viscosity [45]. Moreover, all the DES's temperatures depended on the viscosity profile followed by Arrhenius behavior.

3.4.1.2 Density

Density plays a crucial role in optimizing any process. A recent study by Ge *et al.* showed an increase in density with the addition of glycerol monolaurate (GML) in thymol/GML-based DES [48]. This could be due to the molecular interaction between thymol and GML, which increased the energy dispersion factor by restricting fluid motion. The destiny of the thymol/coumarin was highest among the other five DESs [46]. In LLE, the denser the HDES, the better the pollutant's solubilization. It was stated that a longer alkyl chain of the HBA reduces the density of the HDES due to the presence of HBD. Unlike this, another observed the effect of HBA on density [45]. With the increase in lactic acid ratio, the density of the natural DESs increased. In addition, temperature effects on density are essential for industrial applications. Similar to viscosity, a reduction in density was observed with an increase in temperature [45].

3.4.1.3 Refractive Index

The refractive index (RI) shows the purity and concentration of the compounds of various liquid and solid samples. Ge *et al.* study suggests that RI decreases in thymol/GML DES with increased GML [48]. This can be explained by the fact that the precursors greatly influence the properties of the DES.

3.4.1.4 Surface Tension and Contact Angle

At the surface of a liquid, the cohesion between the molecules denotes surface tension (ST), whereas contact angle indicates the wettability of the surface or material. These are essential for understanding solvents' interfacial behavior and separation application. Al-Bodour *et al.* reported that both carvone/lactic acid (1:1) and cineole/lactic acid (3:1) showed high wettability on glass [45]. The lowest ST were on cineole/lactic acid (3:1), cineole/lactic acid (3:1), and carvone/lactic acid (1:1) on SS-304. These findings show that their surface mobility is higher and their resistance to penetration is lower, which is suitable for separation.

3.4.1.5 *Thermal Property*

One of the critical criteria for HDES is that its melting temperature is visibly lower than that of the precursors. Also, their thermal stability is identified by the thermogravimetric analysis. A better thermal range is preferred for the operational process [45].

3.4.2 Terpene-Based DESs' Applications

Over the years, terpenes have been used for various applications in diverse fields, from medicinal to flavoring agents in food, fragrances, cosmetics, etc. Terpene-based NDES recently gained popularity due to their hydrophobicity. It found numerous applications from LLE, extraction, CO_2 capture, electrochemical application, etc.

A hydrophobic DES, made from menthol and acetic acid, was used as a medium to study electrochemical applications. Ethanolamine was added to it to increase the conductivity of the DES. The electrochemical behavior of hydroxymethyl ferrocene (FcMeOH), an electroactive mediator and reversible electrochemical system, has been studied using cyclic voltammetry. Its coefficient of diffusion, ascertained in DESs, is comparable in magnitude to that found for the most widely utilized IL for electrochemical applications, 1-ethyl-3 methylimidazolium bis(trifluoromethanesulfonyl) imide ([EMIM][TFSI]) [49]. Selenium is widely used as a supplement in the poultry industry. It accumulates in eggs and chicken products and harms humans with inadequate uptake. In a recent study, terpenoid-based HDESs were studied for liquid-liquid microextraction to investigate the presence of selenium in eggs and chicken tissue [50]. Apart from that, Ge *et al.* studied the antimicrobial efficacy and stability of eutectic nano-emulsion made from glycerin monolaurate and thymol-based DES [48]. Fresh pork's shelf life increased when treated with optimized eutectic nano-emulsions (40% thymol/60% GML). This was mainly because the encapsulated essential oil prevented lipid oxidation and microbiological development. Terpene-based DES was investigated for removing contaminants, i.e., cresol from

WW [46]. Terpene DESs, being hydrophobic, greatly influenced the said LLE process. Moreover, the cresol extraction process was greatly influenced by the strong H-bonding. Also, the molar ratio of the DES affected the extraction process due to reduced surface area and higher polarity. In addition, higher contact time increases the LLE process. Another study by Al-Bodour *et al.* studied lactic acid and various terpenes used for CO_2 capture [45]. The low viscosity and density of the studied DES improved CO_2 uptake. At 308.15 K, CO_2 absorption was rapid compared to 298.15 K in cineole/lactic acid (1:1) DES, as viscosity reduced at high temperatures. The selection of HBD and HBA greatly influences the absorption kinetics of CO_2 and solubility in NDES; the primary determinants are molecular interactions, specifically hydrogen bonding and van der Waals forces.

3.5 Some Other Green Solvents

In addition to ILs and DESs discussed previously, some other green solvents are also included in this section. A brief discussion on them can be found below.

3.5.1 Non-Toxic Liquid Polymers

Liquid polymers, namely, polyethylene glycol (PEG) and polypropylene glycol (PPG), offer significant benefits as green solvents. It is well-known that PEG is less-toxic, non-flammable, and biodegradable, which makes it a greener alternative for a series of applications. It is also used as an inhibitor and catalyst. Both PPG and PEG possess comparable properties. Other liquid polymers, e.g., polydimethylsiloxane and polytetrahydrofuran are less biodegradable and may cause risks to water bodies.

3.5.2 Supercritical Fluids

Supercritical carbon dioxide and supercritical water are examples of supercritical fluids. They exist above their critical points, having properties of both gases and liquids. Supercritical carbon dioxide is well-known for its use in polymer processing. It is considered as a green alternative due to its reduced environmental impact and high solubility in polymers. Similarly, supercritical water also provides certain advantages for its applications in green chemistry.

3.5.3 Common Industrial Applications

Research on green solvents is ongoing, with continuously growing alternatives by the scientists. Their sustainability is the key factor before industrial scale-up. A few common industrial applications are summarized in Table 3.9 [51].

Table 3.9 Industrial applications of green solvent.

Industry	Application
Pharmaceutical production	Extraction of valuable drugs and natural products
Chemical synthesis	As reaction media for creating desired chemicals
Material processing	Dissolving and processing polymers, coatings, and other materials
Cleaning and degreasing	Replacing hazardous solvents in cleaning processes

3.6 Conclusion

Environment and energy are the major concerns of today's scientific community. Researchers worldwide are continuously striving to provide sustainable solutions to current energy demand without causing any damage to the environment. The study reveals the potential of ILs and DESs for several multidisciplinary applications with primary focus on CO_2 capture and electrical double layer capacitors. DESs can successfully replace the well-known amines for solvent absorption of post-combustion CO_2. Physisorption is the major advantage with DESs in order to minimize the energy inputs and have a cost-effective process. The discussion also shows how different HBA-HBD combinations can affect the solubility of CO_2 in DESs. Considering all the cons of DESs, further research and development is needed to improvise the properties. It is well understood that there is no "ideal solvent" that is suitable for all the applications without much harmful effects to the environment. The flexibility of designing a DES based on the desired properties facilitates their applications and research. Certainly, significant work on DESs' versatile application and their industrial application can be expected in coming years. The most positive aspect of DESs is their cost-effectiveness compared to ILs. This opens the door to explore more and more DESs with properties matching with ILs.

AI disclosure statement: The author confirms that no artificial intelligence (AI)-generated content or AI-assisted writing tools were used in the preparation of this chapter (including summaries and the abstract). All text, analysis, and conclusions are the result of the author's original work.

References

1. Lei, Z., Chen, B., Koo, Y.-M., MacFarlane, D.R., Introduction: Ionic Liquids. *Chem. Rev.*, 117, 10, 6633–6635, May 2017, doi: 10.1021/acs.chemrev.7b00246.
2. Smith, E.L., Abbott, A.P., Ryder, K.S., Deep Eutectic Solvents (DESs) and Their Applications. *Chem. Rev.*, 114, 21, 11060–11082, Nov. 2014, doi: 10.1021/cr300162p.
3. Tsuda, T. and Hussey, C.L., Electrochemical Applications of Room-Temperature Ionic Liquids. *Electrochem. Soc. Interface*, 16, 1, 42–49, Mar. 2007, doi: 10.1149/2.F05071IF.
4. Fedorov, M.V. and Kornyshev, A.A., Ionic Liquids at Electrified Interfaces. *Chem. Rev.*, 114, 5, 2978–3036, Mar. 2014, doi: 10.1021/cr400374x.
5. Singh, V.V., Nigam, A.K., Batra, A., Boopathi, M., Singh, B., Vijayaraghavan, R., Applications of Ionic Liquids in Electrochemical Sensors and Biosensors. *Int. J. Electrochem.*, 2012, 1–19, 2012, doi: 10.1155/2012/165683.
6. Eftekhari, A., Supercapacitors utilising ionic liquids. *Energy Storage Mater.*, 9, 47–69, Oct. 2017, doi: 10.1016/j.ensm.2017.06.009.
7. Mahanta, U., Venkatesh, R.P., Sujatha, S., Ilangovan, S.A., Banerjee, T., Imidazolium Based Ionic Liquids as Electrolytes for Energy Efficient Electrical Double Layer Capacitor: Insights from Molecular Dynamics and Electrochemical Characterization. *J. Solution Chem.*, 48, 7, 1119–1134, Jul. 2019, doi: 10.1007/s10953-019-00898-8.
8. Mahanta, U., Kundu, D., Venkatesh, R.P., Sujatha, S., Ilangovan, S.A., Banerjee, T., Electrochemical Performance and Molecular Structure of Diluted 1-Alkyl-3-methylimidazolium Tetrafluoroborate Ionic Liquids and Their Mixture as Electrolytes for Double-Layer Capacitors: An Integrated Approach by Electrochemical Characterization and Molecular Dynamics Simulation. *Ind. Eng. Chem. Res.*, 58, 51, 22741–22753, Dec. 2019, doi: 10.1021/acs.iecr.9b04350.
9. Mahanta, U., Koradiya, A., Prasanna Venkatesh, R., Sujatha, S., Ilangovan, S.A., Banerjee, T., Electrochemical performance of 1-butyl-4-methylpyridinium tetrafluoroborate ionic liquid electrolyte for graphene-based double layer capacitor. *Bull. Mater. Sci.*, 44, 75, Mar. 2021, doi: 10.1007/s12034-020-02337-7S.
10. Dutta, A., *et al.*, Examining the Electrochemical Nature of an Ionogel Based on the Ionic Liquid [P 66614][TFSI] and TiO 2 : Synthesis, Characterization, and Quantum Chemical Calculations. *Ind. Eng. Chem. Res.*, 61, 25, 8763–8774, Jun. 2022, doi: 10.1021/acs.iecr.2c00550.
11. Mahanta, U., Ionic liquids and deep eutectic solvents as electrolytes for energy efficient electrical double layer capacitor, PhD Thesis, Indian Institute of Technology Guwahati, Guwahati, 2021.

12. Bedrov, D., Piquemal, J.-P., Borodin, O., MacKerell, A.D., Roux, B., Schröder, C., Molecular Dynamics Simulations of Ionic Liquids and Electrolytes Using Polarizable Force Fields. *Chem. Rev.*, 119, 13, 7940–7995, Jul. 2019, doi: 10.1021/acs.chemrev.8b00763.
13. Endres, F., Physical chemistry of ionic liquids. *PCCP*, 12, 8, 1648, Feb. 2010, doi: 10.1039/c001176m.
14. Shang, D., *et al.*, Ionic liquids in gas separation processing. *Curr. Opin. Green Sustain. Chem.*, 5, 74–81, Jun. 2017, doi: 10.1016/j.cogsc.2017.03.015.
15. Neumann, J.G. and Stassen, H., Anion Effect on Gas Absorption in Imidazolium-Based Ionic Liquids. *J. Chem. Inf. Model.*, 60, 2, 661–666, Feb. 2020, doi: 10.1021/acs.jcim.9b00885.
16. Lapshin, D.N., Jorge, M., Campbell, E.E.B., Sarkisov, L., On competitive gas adsorption and absorption phenomena in thin films of ionic liquids. *J. Mater. Chem. A Mater.*, 8, 23, 11781–11799, Jun. 2020, doi: 10.1039/D0TA03419C.
17. Liu, X. and Turner, C.H., Understanding gas absorption in multivalent ionic liquids *via* solute-solvent interaction analyses. *Chem. Phys. Lett.*, 786, 139204, Jan. 2022, doi: 10.1016/j.cplett.2021.139204.
18. Sosa, J.E., *et al.*, Design of Ionic Liquids for Fluorinated Gas Absorption: COSMO-RS Selection and Solubility Experiments. *Environ. Sci. Technol.*, 56, 9, 5898–5909, May 2022, doi: 10.1021/acs.est.2c00051.
19. Zhao, Y., Gani, R., Afzal, R.M., Zhang, X., Zhang, S., Ionic liquids for absorption and separation of gases: An extensive database and a systematic screening method. *AlChE J.*, 63, 4, 1353–1367, Apr. 2017, doi: 10.1002/aic.15618.
20. Zhang, X., Zhang, S., Bao, D., Huang, Y., Zhang, X., Absorption degree analysis on biogas separation with ionic liquid systems. *Bioresour. Technol.*, 175, 135–141, Jan. 2015, doi: 10.1016/j.biortech.2014.10.048.
21. Greer, A.J., Jacquemin, J., Hardacre, C., Industrial Applications of Ionic Liquids. *Molecules*, 25, 21, 5207, Nov. 2020, doi: 10.3390/molecules25215207.
22. Zhuang, L., *et al.*, Enhancing the solubility of carbamazepine using ionic liquids: An experiment and theoretical calculation. *J. Mol. Struct.*, 1317, 139198, Dec. 2024, doi: 10.1016/j.molstruc.2024.139198.
23. Akhavein, N., *et al.*, Parenteral platforms for tunable, long-acting administration of a highly hydrophobic antiretroviral drug. *Sci. Rep.*, 14, 1, 11573, May 2024, doi: 10.1038/s41598-024-58583-w.
24. Pansuriya, R., *et al.*, Self-healable, stimuli-responsive bio-ionic liquid and sodium alginate conjugated hydrogel with tunable Injectability and mechanical properties for the treatment of breast cancer. *Int. J. Biol. Macromol.*, 277, 134112, Oct. 2024, doi: 10.1016/j.ijbiomac.2024.134112.
25. Pandya, I., *et al.*, 5-Fluoro uracil mediated transformation of surface-active ionic liquid based micellar nanoaggregates into pH-responsive vesicular nanoaggregates as promising drug delivery vehicle. *J. Mol. Liq.*, 409, 125443, Sep. 2024, doi: 10.1016/j.molliq.2024.125443.
26. Pârvulescu, V.I., and Hardacre, C., Catalysis in Ionic Liquids. *Chem. Rev.*, 107, 6, 2615–2665, Jun. 2007, doi: 10.1021/cr050948h.

27. Welton, T., Ionic liquids in catalysis. *Coord. Chem. Rev.*, 248, 21–24, 2459–2477, Dec. 2004, doi: 10.1016/j.ccr.2004.04.015.
28. Tomé, L.I.N., Baião, V., da Silva, W., Brett, C.M.A., Deep eutectic solvents for the production and application of new materials. *Appl. Mater. Today*, 10, 30–50, Mar. 2018, doi: 10.1016/j.apmt.2017.11.005.
29. García, G., Aparicio, S., Ullah, R., Atilhan, M., Deep Eutectic Solvents: Physicochemical Properties and Gas Separation Applications. *Energy Fuels*, 29, 4, 2616–2644, Apr. 2015, doi: 10.1021/ef5028873.
30. Malaquias, J.C., Steichen, M., Thomassey, M., Dale, P.J., Electrodeposition of Cu–In alloys from a choline chloride based deep eutectic solvent for photovoltaic applications. *Electrochim. Acta*, 103, 15–22, Jul. 2013, doi: 10.1016/j.electacta.2013.04.068.
31. Hadj-Kali, M.K., Al-khidir, K.E., Wazeer, I., El-blidi, L., Mulyono, S., AlNashef, I.M., Application of deep eutectic solvents and their individual constituents as surfactants for enhanced oil recovery. *Colloids Surf. A Physicochem. Eng. Asp.*, 487, 221–231, Dec. 2015, doi: 10.1016/j.colsurfa.2015.10.005.
32. Ghareh Bagh, F.S., Shahbaz, K., Mjalli, F.S., Hashim, M.A., AlNashef, I.M., Zinc (II) chloride-based deep eutectic solvents for application as electrolytes: Preparation and characterization. *J. Mol. Liq.*, 204, 76–83, Apr. 2015, doi: 10.1016/j.molliq.2015.01.025.
33. Farajzadeh, M.A., Afshar Mogaddam, M.R., Feriduni, B., Simultaneous synthesis of a deep eutectic solvent and its application in liquid–liquid microextraction of polycyclic aromatic hydrocarbons from aqueous samples. *RSC Adv.*, 6, 53, 47990–47996, 2016, doi: 10.1039/C6RA04103E.
34. Dehury, P., Mahanta, U., Banerjee, T., Comprehensive Assessment on the Use of Boron Nitride-Based Nanofluids Comprising Eutectic Mixtures of Diphenyl Ether and Menthol for Enhanced Thermal Media. *ACS Sustain. Chem. Eng.*, 8, 38, 14595–14604, Sep. 2020, doi: 10.1021/acssuschemeng.0c05648.
35. Dehury, P., Mahanta, U., Singh, R., Banerjee, T., Potential of deep eutectic solvent based nanofluids as a new generation heat transfer media. *J. Mol. Liq.*, 379, 121700, Jun. 2023, doi: 10.1016/j.molliq.2023.121700.
36. Arrhenius, S., XXXI. *On the influence of carbonic acid in the air upon the temperature of the ground. Lond. Edinb. Dublin Philos. Mag. J. Sci.*, 41, 251, 237–276, Apr. 1896, doi: 10.1080/14786449608620846.
37. Hekmatmehr, H., *et al.*, Carbon capture technologies: A review on technology readiness level. *Fuel*, 363, 130898, May 2024, doi: 10.1016/j.fuel.2024.130898.
38. Siani, G., *et al.*, Physical absorption of CO_2 in betaine/carboxylic acid-based Natural Deep Eutectic Solvents. *J. Mol. Liq.*, 315, 113708, Oct. 2020, doi: 10.1016/j.molliq.2020.113708.
39. Li, X., Hou, M., Han, B., Wang, X., Zou, L., Solubility of CO 2 in a Choline Chloride + Urea Eutectic Mixture. *J. Chem. Eng. Data*, 53, 2, 548–550, Feb. 2008, doi: 10.1021/je700638u.
40. Karami, B., Ghaemi, A., Shahhosseini, S., Eco-Friendly Deep Eutectic Solvents Blended with Diethanolamine Solution for Postcombustion

CO_2 Capture. *Energy Fuels*, 36, 2, 945–957, Jan. 2022, doi: 10.1021/acs.energyfuels.1c03002.

41. Brahmkshatriya, P.P. and Brahmkshatriya, P.S., Terpenes: Chemistry, biological role, and therapeutic applications, in: *Natural Products: Phytochemistry, Botany and Metabolism of Alkaloids, Phenolics and Terpenes*, Springer Berlin Heidelberg, pp. 2665–2691, 2013, doi: 10.1007/978-3-642-22144-6_120.
42. Masyita, A. *et al.*, Terpenes and terpenoids as main bioactive compounds of essential oils, their roles in human health and potential application as natural food preservatives. *Food Chem. X*, 13, 100217, Mar. 2022, doi: 10.1016/j.fochx.2022.100217.
43. Kapadia, P., Newell, A.S., Cunningham, J., Roberts, M.R., Hardy, J.G., Extraction of High-Value Chemicals from Plants for Technical and Medical Applications. *Int. J. Mol. Sci.*, 23, 18, 10334, Sep. 2022, doi: 10.3390/ijms231810334.
44. Mosquera, M.E.G. *et al.*, Terpenes and Terpenoids: Building Blocks to Produce Biopolymers. *Sustain. Chem.*, 2, 3, 467–492, Aug. 2021, doi: 10.3390/suschem2030026.
45. Al-Bodour, A., Alomari, N., Aparicio, S., Atilhan, M., A comprehensive study on carbon capture potential of lactic acid based deep eutectic solvents at wide process conditions. *J. Mol. Liq.*, 390, 123114, Nov. 2023, doi: 10.1016/j.molliq.2023.123114.
46. Wazeer, I., Hizaddin, H.F., El blidi, L., Mulyono, S., Hashim, M.A., Hadj-Kali, M. K., Removal of cresol contaminants from aqueous media using hydrophobic deep eutectic solvents. *J. Mol. Liq.*, 407, 125082, Aug. 2024, doi: 10.1016/j.molliq.2024.125082.
47. Abranches, D.O. and Coutinho, J.A.P., Type V deep eutectic solvents: Design and applications. *Curr. Opin. Green Sustain. Chem.*, 35, 100612, Jun. 2022, doi: 10.1016/j.cogsc.2022.100612.
48. Ge, Y. *et al.*, Formation, stability, and antimicrobial efficacy of eutectic nanoemulsions containing thymol and glycerin monolaurate. *Food Chem.*, 453, 139689, Sep. 2024, doi: 10.1016/j.foodchem.2024.139689.
49. Chaabene, N., Ngo, K., Turmine, M., Vivier, V., New hydrophobic deep eutectic solvent for electrochemical applications. *J. Mol. Liq.*, 319, 114198, Dec. 2020, doi: 10.1016/j.molliq.2020.114198.
50. Shishov, A., Krekhova, F., Bulatov, A., Deep eutectic solvent based on thymol and menthol for microextraction and sensitive UV-Vis determination of total selenium in eggs and chicken tissue. *J. Food Compos. Anal.*, 126, 105875, Feb. 2024, doi: 10.1016/j.jfca.2023.105875.
51. Clarke, C.J., Tu, W.-C., Levers, O., Bröhl, A., Hallett, J.P., Green and Sustainable Solvents in Chemical Processes. *Chem. Rev.*, 118, 2, 747–800, Jan. 2018, doi: 10.1021/acs.chemrev.7b00571.

4

Intensification Strategies in Green Engineering

Hanisha R., Dhanashree T.D., Sangeetha Priya R.B., Subanu M. and Gopinath M.*

Department of Biotechnology, Veltech Hightech Dr. Rangarajan Dr. Sakunthala Engineering College, Chennai, India

Abstract

The concept of green technology implies the application of eco-friendly materials and procedures that aid in reducing emissions and the preservation of natural resources. The ever-growing demand for resources and products necessitates sustainable and environmentally conscious practices. Green engineering offers a solution by focusing on designing processes and products with minimal environmental impact. In green engineering, intensification strategies refer to a range of methods used to improve the sustainability and effectiveness of green technology. These tactics could involve using resources as efficiently as possible, increasing energy efficiency, cutting down on waste production, and utilizing renewable energy sources. Adopting eco-friendly materials, encouraging sustainable design techniques, and putting lifetime evaluations into practice are other essential elements of intensification tactics in green engineering. This work explores various intensification strategies employed in green engineering. These strategies aim to improve process efficiency, reduce energy consumption, and minimize waste generation. Examples include enhancing transport processes through microreactors and static mixers, integrating reaction and separation steps, and utilizing alternative energy sources like photocatalysis and electrocatalysis. The adoption of these strategies leads to several benefits. Intensified processes are more compact, require less energy, and generate less waste. This translates to reduced environmental impact, lower operational costs, and improved product quality.

Keywords: Energy efficiency, renewable, waste minimization, sustainability, green engineering, miniaturization

**Corresponding author*: mpgopinath1980@gmail.com

Senthilkumar Rathnasamy, Vivek Rangarajan and Umile Gianfranco Spizzirri (eds.) Green Engineering and Sustainable Technology: Tools and Techniques, (103–134) © 2026 Scrivener Publishing LLC

4.1 Introduction

Green engineering is an innovative approach focused on designing, developing, and implementing products and processes that are environmentally responsible and sustainable. It integrates principles from various engineering disciplines to minimize environmental impact, promote efficiency, and ensure economic viability: meeting current needs without compromising future generations by using resources responsibly; considering the entire lifecycle of products and processes to mitigate environmental impacts [1]; reducing energy consumption through innovative designs and renewable energy sources, minimizing waste generation and promoting recycling and reuse; preventing pollution at the source by using non-toxic materials and reducing emissions; ensuring cost-effective solutions that provide economic benefits; developing solar, wind, and other renewable energy systems; and constructing energy-efficient and sustainable buildings minimizes exposure to toxins and improves public health [2].

Green engineering is essential for addressing global environmental challenges and creating a sustainable future. Green engineering is an exciting field that focuses on designing and developing processes and products with a minimized environmental impact. It is all about creating solutions that are not only effective but also sustainable for our planet. This ensures that environmental impacts are minimized at every stage. This helps identify areas for improvement and promotes sustainable practices. Green engineering promotes the efficient use of resources and explores the use of renewable resources whenever possible [3]. This protects human health and the environment. By reducing pollution and promoting sustainability, green engineering helps protect our environment and resources for future generations. Designing procedures and goods with the intention of reducing their negative effects on the environment and advancing sustainability is known as green engineering. Using intensification tactics, which maximize resource efficiency and minimize waste creation, is a crucial component of green engineering. These tactics concentrate on raising process productivity while reducing energy, water, and raw material consumption. Process integration, optimization, and innovation are examples of green engineering intensification techniques. While process optimization focuses on enhancing the performance of current processes through the application of cutting-edge technology and techniques, process integration combines several unit operations to create more efficient processes [4].

Creating new methods and tools that lessen their negative effects on the environment and increase sustainability as a whole is considered innovation. The goal of the field of "green engineering" is to create, develop, and apply long-term fixes for environmental problems. The application of intensification tactics, which seek to optimize effectiveness and reduce resource consumption in engineering processes, is a crucial component of green engineering. The present study aims to investigate diverse methodologies and implementations of intensification strategies in green engineering, emphasizing their possible advantages and consequences for sustainable development [5]. This may work toward a future that is more resource- and environmentally-conscious by knowing how these techniques can be used in various technical contexts. One key area of advancement in intensification strategies is the integration of process intensification (PI) techniques, such as microreactors, membrane technologies, and continuous processing, into traditional engineering practices. These techniques enable the efficient use of resources, reducing the overall environmental footprint of industrial processes while improving productivity and cost-effectiveness. Additionally, advances in digital technologies, such as artificial intelligence and machine learning, have enabled the optimization of processes in real time, further enhancing the sustainability of green engineering practices [6].

Furthermore, the implementation of circular economy principles in intensification strategies has emerged as a key trend in promoting sustainability in green engineering. By closing the loop on resource use and waste generation, circular economy approaches help minimize environmental impact and maximize resource efficiency, contributing to a more sustainable future for industries worldwide [7]. Overall, intensification strategies in green engineering play a crucial role in helping industries reduce their environmental footprint and move toward a more sustainable future. By adopting these strategies, companies can improve their competitiveness, reduce costs, and contribute to a cleaner and healthier environment for future generations.

4.2 Process Intensification in Green Engineering

PI in green engineering refers to the concept of designing and optimizing chemical processes to minimize waste, energy consumption, and environmental impact. This approach focuses on increasing efficiency, reducing the use of resources, and incorporating sustainable practices to promote environmentally friendly manufacturing processes. There are several ways in which PI can be implemented in green engineering. Miniaturization involves designing smaller and more compact reactors and equipment to reduce the number of raw materials and energy required for a process [8]. By integrating different processes or unit operations into one compact system, overall efficiency can be increased and waste generation minimized.

Novel reactor designs are an innovation in reactor design, such as microchannel reactors or membrane reactors, which can improve heat and mass transfer rates, leading to more sustainable processes. *In situ* separation techniques that involves the separation processes that are incorporated directly into the reaction step can reduce energy consumption and waste generation. Implementing continuous processing instead of batch processing can lead to significant reductions in energy consumption and waste generation. Use of alternative green solvents that substituting traditional solvents with environmentally friendly alternatives can reduce the environmental impact of chemical processes. Designing processes that generate minimal waste and incorporate recycling of by-products can help promote sustainability. By incorporating these principles of PI in green engineering, companies can improve their environmental footprint, reduce their reliance on fossil fuels, and contribute to a more sustainable future [9].

4.3 Importance of Green Engineering Benefits

PI is a game-changer in green engineering, offering a powerful toolbox for designing more sustainable and environmentally friendly processes. Green engineering is critical in addressing today's pressing environmental, economic, and social challenges. PI plays a critical role in green engineering by promoting sustainable practices, reducing environmental impact, and enhancing process efficiency [10]. This paves the way for a greener future for the chemical process industry and beyond.

4.3.1 Reduced Environmental Impact

PI techniques often involve smaller reaction vessels and more efficient use of materials. This translates to less waste generation, both in terms of unused reactants and unwanted by-products. By optimizing processes and reducing energy consumption, PI helps minimize air and water pollution associated with traditional manufacturing methods. It promotes efficient use of raw materials and water throughout the process. This helps conserve precious resources for future generations [11]. Green engineering focuses on preventing pollution at the source, reducing the release of harmful substances into the air, water, and soil. This contributes to cleaner ecosystems and improved biodiversity. By optimizing the use of natural resources and encouraging the use of renewable resources, green engineering helps in conserving vital resources like water, minerals, and forests.

4.3.2 Enhanced Process Efficiency

Green engineering promotes practices that ensure long-term sustainability, enabling societies to meet present needs without compromising future generations' ability to meet their own. By fostering sustainable practices, green engineering helps communities and industries become more resilient to environmental changes and disruptions. Many PI techniques promote faster reaction rates, improved heat and mass transfer, and efficient use of energy [12]. This translates to significant reductions in energy consumption, leading to lower operational costs and a smaller carbon footprint. That allows for miniaturization of reaction vessels and integration of multiple processing steps. This leads to smaller, more compact equipment footprints, reducing space requirements and associated costs. Intensified processes often operate at faster rates due to improved mixing, heating, and mass transfer. This translates to faster production times and increased output for a given amount of time [13].

4.3.3 Economic and Societal Benefits

Green engineering encourages the development of new technologies and innovative solutions which is shown in Figure 4.1 that can lead to

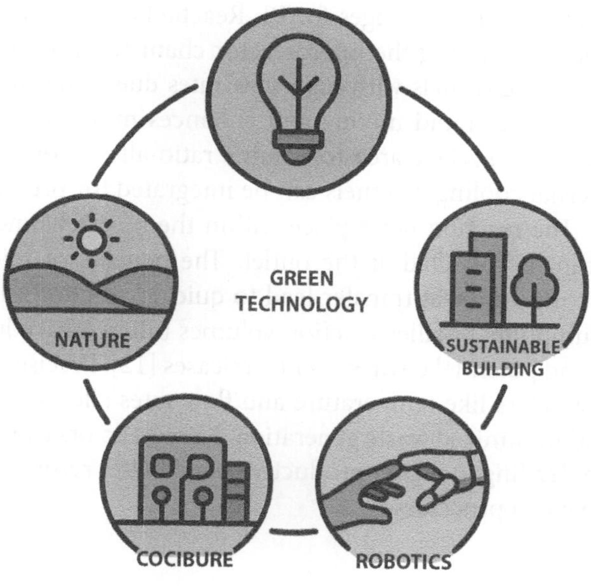

Figure 4.1 Highlights the role of green technology in driving innovation and advancement.

breakthroughs in various industries. The integration of principles from different engineering disciplines promotes collaboration and knowledge sharing, leading to more holistic and effective solutions. Reduced energy consumption, smaller equipment, and minimized waste disposal costs translate to significant economic benefits for industries adopting PI techniques. Intensified processes can offer more precise control over reaction conditions, leading to improved product quality and consistency [14]. By promoting resource efficiency and environmental protection, PI contributes to achieving sustainable development goals for a healthier planet.

4.4 Examples of Process Intensification Techniques

PI plays a crucial role in green engineering by promoting sustainable practices and reducing environmental impact. Microreactors, ultrasonic processing, and microwave heating are the techniques that are involved in PI techniques.

4.4.1 Microreactors

Microreactors are miniaturized reactors with channel sizes typically ranging from 10 to 500 micrometers. Their high surface area-to-volume ratio offers significant advantages for PI. Reactants are precisely metered and mixed before entering the microreactor channels. The mixture flows through the microchannels with high flow rates due to the small dimensions. This promotes rapid mixing and enhances mass transfer between reactants. The high surface area-to-volume ratio allows for efficient heat transfer. External cooling channels can be integrated for precise temperature control. The reaction takes place within the microchannels, and the product stream is collected at the outlet. The benefits of microreactors are faster mixing and heat transfer lead to quicker reactions, minimizing energy consumption. Smaller reaction volumes inherently reduce the risk of accidents and potential environmental releases [12]. Precise control over reaction parameters like temperature and flow rates allows for optimized processes and minimized waste generation. Microreactors can be operated continuously, leading to higher productivity and better resource utilization compared to batch processes.

4.4.2 Ultrasonic Processing

Ultrasonics utilizes high-frequency sound waves (above 20 kHz) to create various effects in liquids, including cavitation, acoustic streaming, and mechanical effects. An electrical signal is generated at the desired frequency. The signal is converted into mechanical vibrations (sound waves) by a piezoelectric transducer. The sound waves propagate through the reaction medium (liquid). These waves can create rapid formation and collapse of bubbles, generating intense localized forces and heat for promoting reactions, emulsification, or cell disruption. Bulk flow of the liquid occurs in the direction of wave propagation, which enhances mixing and mass transfer [16]. Depending on the material properties, vibrations can cause particle disruption, erosion, or degassing. Enhanced mixing and mass transfer due to cavitation and streaming can accelerate reactions, reducing processing times and energy consumption. Cavitation is a powerful tool for cleaning surfaces without harsh chemicals, minimizing waste generation. Ultrasonics can be used for extraction, homogenization, and emulsification, potentially replacing energy-intensive or solvent-based processes.

4.4.3 Microwave Heating

Microwave heating directly heats polar molecules within a material. This volumetric heating mechanism penetrates throughout the material, providing uniform and rapid heating compared to conventional methods. The reaction mixture is placed within a microwave cavity. Microwaves are directed into the cavity, causing polar molecules in the mixture to vibrate and generate heat. The volumetric heating ensures even heat distribution throughout the material, eliminating hot and cold spots [15]. The reaction proceeds due to the microwave-induced heating, and the product is collected after the process. Faster and more efficient heating compared to conventional methods leads to significant energy savings. Selective heating of specific molecules allows for better control over reaction rates and product quality. Microwave heating can be used in solvent-free or low-solvent conditions, minimizing waste generation and solvent emissions. Rapid and uniform heating translates to faster reaction times and higher process throughput.

These are just a few examples of PI techniques used in green technology. By implementing these strategies, engineers can design more sustainable and efficient chemical processes, paving the way for a greener future. Remember, the selection of the most suitable PI technique depends on the specific process and desired outcomes [17].

4.5 Frequently Employed Sustainable Extraction Techniques

4.5.1 Microwave-Assisted Extraction

Microwave-assisted extraction (MAE) is an advanced technique utilized to efficiently extract bioactive compounds, essential oils, and other valuable substances from various matrices, such as plants, foods, and environmental samples. MAE is a rising star in green technology engineering, offering a sustainable and efficient approach to extracting valuable components from various materials. MAE utilizes microwave radiation to accelerate the extraction process. Microwaves interact with polar molecules within the sample, causing them to vibrate and generate heat. This internal heating mechanism offers several advantages over traditional extraction methods [18]. Compared to conventional heating methods, MAE significantly reduces extraction times, leading to lower energy consumption and improved process efficiency. Microwaves can penetrate deeper into the sample matrix, promoting better solvent-analytic interaction and potentially increasing the yield of extracted components. MAE often allows for the use of less solvent or even solvent-free extraction methods in some cases. This minimizes waste generation and solvent-related environmental impact. The ability of microwaves to target specific polar molecules can be advantageous. This selectivity can improve the quality and purity of the extracted components.

A traditional method that can be time-consuming requires large solvent volumes and has limited selectivity. It shares some similarities with MAE but may have higher equipment costs and requires careful control to avoid degrading heat-sensitive compounds. Choosing the right solvent is crucial for efficient extraction and compatibility with microwave heating. Overheating can damage heat-sensitive compounds [19]. Careful control of temperature and exposure time is necessary. MAE processes developed in the laboratory might need adjustments for efficient scale-up to industrial applications. MAE presents a promising approach for sustainable extraction in green technology engineering. Its ability to reduce processing times, minimize solvent usage, and potentially enhance extraction yields makes it a valuable tool for a greener future.

4.5.1.1 Microwave-Assisted Extraction Principle

Microwaves cause polar molecules in the sample to oscillate, generating heat through dielectric heating. This Figure 4.2 shows the stages in MAE

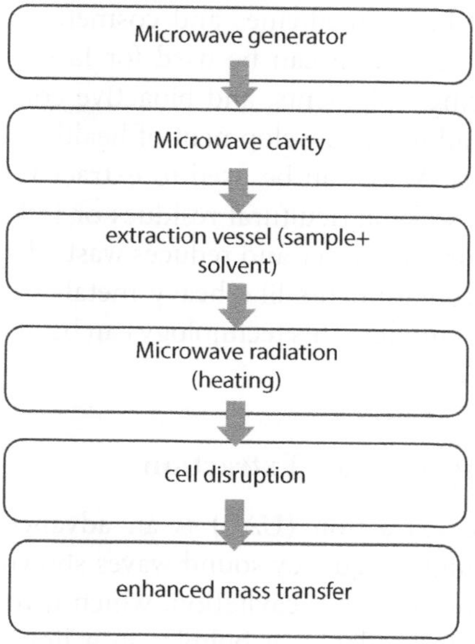

Figure 4.2 Stages of microwave-assisted extraction.

for rapid and uniform heating disrupts cell walls, enhancing the release of target compounds. MAE can be performed with a variety of solvents, including water, ethanol, and other organic solvents [20].

The choice of solvent depends on the solubility of the target compounds. MAE is known for its high extraction efficiency, reduced extraction times, and lower solvent consumption compared to conventional extraction methods. MAE utilizes microwave radiation to accelerate the extraction process. Microwaves interact with polar molecules within the sample, causing them to vibrate and generate heat. Microwaves can penetrate deeper into the sample matrix, promoting better solvent-analyte interaction and potentially increasing the yield of extracted components [21]. MAE often allows for the use of less solvent or even solvent-free extraction methods in some cases. This minimizes waste generation and solvent-related environmental impact. The ability of microwaves to target specific polar molecules can be advantageous. This selectivity can improve the quality and purity of the extracted components.

4.5.1.2 Applications of MAE

MAE is used to extract valuable compounds like essential oils, alkaloids, and phenolics from plant materials. This is particularly beneficial for natural product-based medicines and cosmetics with reduced environmental impact [22]. MAE can be used for faster and more efficient extraction of flavorings, pigments, and bioactive compounds from food sources. This can lead to the development of healthier and more concentrated food products. MAE can be used to extract valuable components from waste materials like agricultural residues or industrial by-products. This promotes resource recovery and reduces waste disposal. MAE can be applied to extract contaminants like heavy metals or organic pollutants from soil and water samples. This technology can be used for environmental cleanup efforts [23].

4.5.2 Ultrasound-Assisted Extraction

Ultrasound-assisted extraction (UAE) is an advanced extraction technique that utilizes high-frequency sound waves shown in Figure 4.3. The key mechanism behind UAE is cavitation, which involves the formation, growth, and implosive collapse of bubbles in a liquid medium. UAE is another powerful tool in the green technology engineer's toolbox, offering a sustainable and efficient approach to extracting valuable components from various materials. Cavitation and improved mixing lead to increased access of the solvent to the desired compounds, potentially leading to higher extraction yields [24]. Faster mass transfer and improved solvent-analyte interaction translate to shorter extraction times compared to conventional methods, minimizing energy consumption. Choosing the right solvent is crucial for efficient extraction and compatibility with the effects of ultrasound waves. Power, frequency, and processing time of the ultrasound waves need to be optimized for each specific application to achieve maximum efficiency. UAE presents a valuable technique for sustainable extraction in green technology engineering. Its ability to enhance extraction efficiency, reduce processing times, and potentially minimize solvent usage makes it a key player in developing a greener future [25].

4.5.2.1 Mechanism of Ultrasonic Technology

UAE utilizes high-frequency sound waves (above human hearing, typically >20 kHz) to enhance the extraction process, creates the high-frequency electrical signal, and converts the electrical signal into mechanical vibrations (sound waves). These sound waves interact with the solvent and sample matrix, creating several effects that promote efficient extraction. This is typically made of a piezoelectric material that vibrates when an electric

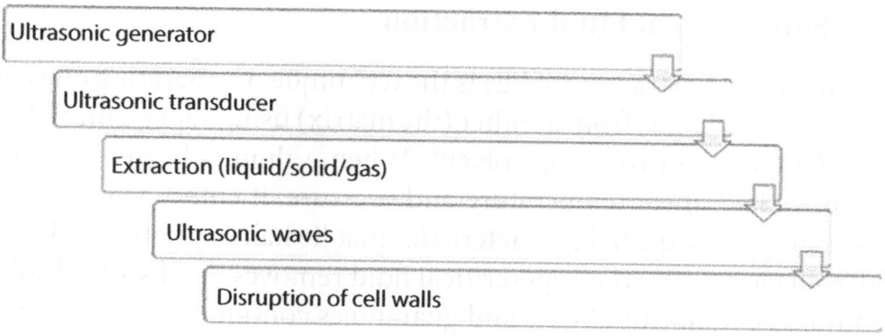

Figure 4.3 Stages of ultrasound-assisted extraction.

field is applied. The medium through which the sound waves travel can be a liquid, solid, or gas depending on the application. The ultrasonic generator produces a high-frequency electrical signal. The transducer converts this electrical signal into mechanical vibrations (sound waves) at the same frequency [26]. These vibrations are transmitted into the medium. High-pressure bubbles form and rapidly collapse in the solvent, generating intense localized forces and heat. This can disrupt cell walls, promote mass transfer, and enhance solvent-analyte interaction. The sound waves create a bulk flow of the solvent within the extraction vessel, improving mixing and bringing fresh solvent into contact with the sample continuously. The vibrations can cause particle disruption, erosion, and degassing depending on the material properties, further aiding in the extraction process [27].

4.5.2.2 *Applications of Ultrasound-Assisted Extraction*

UAE is used to extract valuable compounds like essential oils, alkaloids, and phenolics from plant materials. This is particularly beneficial for natural product-based medicines and cosmetics with reduced environmental impact. UAE can be used to enhance the extraction of biofuels like biodiesel from biomass sources. This promotes the use of renewable and sustainable energy sources [15]. UAE can be applied to extract and recover valuable metals or organic compounds from contaminated soil or water samples, aiding in environmental cleanup efforts. Ultrasonic technology has a wide range of applications across various industries due to the diverse effects that it can produce. Ultrasonic welding uses high-frequency vibrations and pressure to create strong bonds between similar or dissimilar materials. Ultrasonic technology can be used for homogenization, emulsification, and degassing of various materials [28].

4.5.3 Supercritical Fluid Extraction

Supercritical fluid extraction (SFE) is the technique of separating one component (the extractant) from another (the matrix) using supercritical fluids, namely, CO_2 as the extracting solvent. When a fluid is heated and pressured above its critical temperature and pressure, it enters a supercritical state, which gives it special characteristics that lie halfway between those of liquids and gases [29]. The supercritical fluid removes the dissolved solute. It minimizes equilibrium limits and guarantees continuous solute removal through efficient fluid flow. The fluid containing the solute travels through a separator or depressurization chamber. As a result of decreased pressure, the solute becomes less soluble and precipitates. The supercritical fluid can be recycled once the solute has been collected. Solubility parameters are determined by the interactions of solute and solvent molecules.

4.5.3.1 *Mechanism of Supercritical Fluid Extraction*

CO_2 is the most effective plant extraction solvent. Extraction conditions for supercritical CO_2 exceed the critical temperature of 31°C and the critical pressure of 74 bars. Supercritical fluids are extremely compressed gases that exhibit an unusual combination of gas and liquid properties. Supercritical fluids can cause reactions that are difficult or impossible to achieve in normal solvents. It is a quick operation that takes between 10 and 60 minutes to finish. The solid matrix is penetrated by the supercritical fluid. The supercritical fluid's strong solvent power and diffusivity cause solutes to dissolve in it. The supercritical fluid receives solutes that have been liberated from the solid matrix [30]. The fluid's capacity to lessen the intermolecular forces holding the solute to the matrix makes this possible.

The supercritical fluid removes the dissolved solute and minimizes equilibrium limits and guarantees continuous solute removal through efficient fluid flow. The fluid containing the solute travels through a separator or depressurization chamber. As a result of decreased pressure, the solute becomes less soluble and precipitates. The supercritical fluid can be recycled once the solute has been collected. Solubility parameters are determined by the interactions of solute and solvent molecules. Understanding the system's phase behavior is critical for optimizing extraction parameters while the rate of extraction is determined by the mass transfer rates and equilibrium solubility. Continuous flow systems maintain a dynamic equilibrium in which the solute is constantly withdrawn and replaced [31].

4.5.3.2 Principle of Supercritical Fluid Extraction (SFE)

SFE offers a compelling alternative to traditional solvent–based methods for isolating fat-soluble vitamins. The benefits of SFE include significantly lower usage of organic solvents, the prevention of oxygen exposure, and the ability to operate at reduced temperatures. Modern SFE techniques provide shorter extraction times, potentially greater selectivity, and higher sample throughput, thanks to the availability of automated systems, when compared to conventional solvent extraction methods [32]. SFE process is implemented in the ideal circumstances for utilizing SC-CO_2 to extract oleoresins, emphasizing the impact of variables such as temperature, pressure, and co-solvents on extraction yields and quality. Plants contain a wide range of secondary metabolites called phenolic compounds, which are recognized for their antibacterial, anti-inflammatory, and antioxidant properties. SFE shows the effectiveness of SC-CO_2 in phenolic compound extraction from various plant matrices, with a focus on the function of process variables and the possible application of co-solvents such as ethanol to improve extraction efficiency [33].

4.5.3.3 Application and Benefits of Supercritical Fluid Extraction

SFE separates bioactive components from plants, resulting in purer and more effective pharmaceutical and nutraceutical products. SFE, aids in the development of medication formulations with increased bioavailability and stability. SFE removes organic contaminants from soil and trash. It is used to analyze residual solvents in products and ensure compliance with safety regulations, extracting natural oils, vitamins, and fragrances for use in cosmetics and skincare products. SFE enables the isolation of bioactive compounds such as squalene and tocopherols, which are widely used in anti-aging and moisturizing formulations. SFE is used to prepare samples for analysis in chromatography and spectroscopy [34].

4.5.4 Plasma-Assisted Extraction

Plasma-assisted extraction (PAE) is a new technology that uses plasma's unique properties to improve the extraction of bioactive compounds from various matrices. Plasma, also known as the fourth state of matter, is a highly ionized gas consisting of ions, electrons, and neutral particles [35]. Reactive species cause structural damage to the cell walls and facilitate the release of intracellular phytochemicals into the extraction media, hence inducing cellular disruption, improve the process of extracting essential oils that are prized for their aromatic and restorative properties from herbs and spices, and make it easier to extract naturally occurring pigments used in the food and cosmetic industries.

4.5.4.1 Principle of Plasma Technology

Figure 4.4 explains the mechanism of plasma-assisted technology where, initially, plasma is formed by applying an electric field to a gas, which ionizes the gas molecules and produces a mixture of charged particles (electrons and ions) and neutral species. Plasma is commonly generated through dielectric barrier discharge, microwave discharge, and glow discharge. Air, argon, nitrogen, or a combination of these gases is commonly used, and each provides a unique reactive species based on its ionization properties [20].

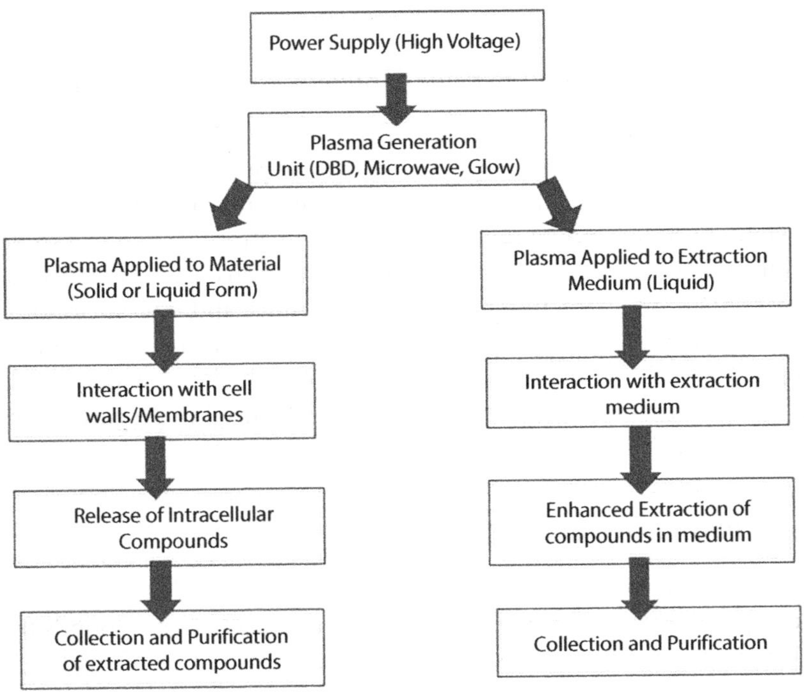

Figure 4.4 Plasma-assisted extraction mechanism.

Reactive species in plasma, such as free radicals, ions, and excited molecules, interact with the material's cell walls and membranes, causing structural damage and allowing intracellular contents to be released. The extraction process can be improved by directly applying plasma treatment to the substance or medium. The isolated chemicals are removed from the media following plasma treatment. Depending on the application, additional purification procedures could be needed to separate the particular chemicals of interest.

4.5.4.2 Mechanism of Plasma-Assisted Green Engineering Extraction (PAE)

Reactive oxygen species (ROS) and reactive nitrogen species (RNS) produced by cold plasma interact with plant cell walls and membranes in ways that include hydroxyl radicals, nitric oxide, and ozone. Reactive species cause structural damage to the cell walls and facilitate the release of intracellular phytochemicals into the extraction media, hence inducing cellular disruption. It effectively extracts flavonoids and phenolic compounds, two types of antioxidants important for their health-promoting qualities, from a variety of plant sources. It improves the process of extracting essential oils that are prized for their aromatic and restorative properties from herbs and spices and makes it easier to extract naturally occurring pigments used in the food and cosmetic industries, such as carotenoids and chlorophylls [36].

4.5.4.3 Applications and Benefits of PAE

Plasma generates ROS and RNS, which interact with the target material. Cellular disruption occurs when these reactive species break down cell walls and membranes, allowing intracellular compounds to be released. Plasma-induced changes improve the mass transfer of bioactive compounds from the material into the extraction medium [37]. PAE operates at low temperatures, which protects the integrity and bioactivity of sensitive compounds. It reduces or eliminates the need for organic solvents, making the process eco-friendlier. When compared to traditional methods, PAE often produces shorter extraction times and higher yields. The process uses less energy, which contributes to overall sustainability.

4.6 Comparative Analysis of Extraction Methods

4.6.1 Comparison of Extraction Mechanism

4.6.1.1 Microwave-Assisted Extraction

Table 4.1 shows the different types of extraction methods MAE uses microwave energy to heat the solvent and plant matrix. Microwaves penetrate the material, rapidly heating and expanding the cell walls. The thermal effects accelerate the diffusion of solutes into the solvent, resulting in faster extraction, and provide uniform and rapid heating, significantly reducing extraction time. High extraction yields are achieved as a result of effective cell disruption and enhanced mass transfer [38].

4.6.1.2 Ultrasound-Assisted Extraction

UAE uses ultrasonic vibrations to generate microbubbles in the solution. Collapsing bubbles cause high pressure and temperature locally. Mechanical actions break down cell walls and increase the solvent's penetration into

Table 4.1 Comparison between the different extraction process.

Features	**MAE (microwave assisted)**	**UAE (ultrasound assisted)**	**SFE (superficial assisted)**	**PAE (plasma assisted)**
Principle	Microwave heating and dielectric heating	Cavitation and mechanical disruption	Solvent properties in superficial state	Reactive species from plasma
Temperature range	Moderate to high	Low to moderate	Low to moderate	Low
Solvent usage	Reduced but depends on material	Moderate	Minimal or none	Minimal or none
Processing time	Short	Short to moderate	moderate to long	Short to moderate
Energy consumption	Moderate to high	-	-	-
References	[24, 23]	[25, 26]	[31, 33]	[44, 46]

the plant matrix; operate at lower temperatures, protecting heat-sensitive chemicals; improve mass transfer and solute diffusion, resulting in higher extraction rates and yields; are suitable for a variety of solvents and materials. High-intensity ultrasound can destroy sensitive bioactive molecules [39].

4.6.1.3 *Supercritical Fluid Extraction*

SFE employs supercritical fluids, usually CO_2, that have both liquid and gas properties above their critical temperature and pressure. The supercritical fluid dissolves solutes like a liquid but diffuses through materials like a gas, allowing for selective compound extraction using pressure and temperature adjustments, and uses CO_2, which is non-toxic, non-flammable, and easily removed from the finished product [40].

4.6.1.4 *Plasma-Assisted Extraction*

PAE generates plasma from an electrical discharge, which contains reactive species like radicals, ions, and UV photons. These reactive species interact with the target material, causing cell walls and membranes to break down and release bioactive chemicals. It operates at low temperatures, making it ideal for heat-sensitive materials, reduces or eliminates the need for organic solvents, which promotes green chemistry principles, and improves extraction efficiency by disrupting cells and transferring mass more effectively [41].

4.6.2 Yield Efficiency and Cost-Effectiveness of Extraction Methods

4.6.2.1 *Microwave-Assisted Extraction (MAE)*

Table 4.2 shows the yield and cost efficiency of different extraction process, in MAE it uses microwave energy to heat solvents in contact with the sample, which improves chemical extraction. MAE demonstrates high extraction efficiency in a significantly shorter time compared to previous procedures. This is especially useful for thermolabile compounds because it limits exposure to high temperatures. Polyphenols, essential oils, and alkaloids have been shown in studies to improve yield significantly. Microwave equipment requires an initial expenditure. Operating expenses are reduced due to shorter extraction times and minimized solvent usage, enabling

Table 4.2 Yield and cost analysis between the different extraction process.

Extraction method	Yield efficiency	Cost-effectiveness	References
MAE (microwave assisted)	High efficiency specially for thermolabile compounds	Moderate to high initial cost	[21, 23]
UAE (ultrasound assisted)	High yield for a wide range of compounds and short extraction time	Low capital and operational cost	[15, 27]
SFE (superficial fluid)	High selectivity and purity	High initial and operational cost	[32, 35]
PAE (plasma assisted)	Promising for heat sensitive and hydrophobic compounds	Moderate to high cost	[39, 41]

scalability for industrial applications and offsetting initial start-up costs through long-term operational savings [42].

4.6.2.2 *Ultrasound-Assisted Extraction (UAE)*

UAE uses ultrasonic vibrations to induce cavitation in the solvent, which increases mass transfer and disrupts cell walls. Flavonoids, carotenoids, and terpenes are among the beneficial substances that can be effectively extracted. It typically delivers excellent yields with shorter extraction periods and lower temperatures and lower capital investment than MAE and SFE. It reduces solvent and energy usage, making it cost-effective for both small- and large-scale applications. Simplified equipment and operational processes help to reduce maintenance expenses [43].

4.6.2.3 *Supercritical Fluid Extraction (SFE)*

SFE uses CO_2 to extract chemicals at temperatures and pressures higher than their critical values, ideal for extracting non-polar substances like lipids, essential oils, and cannabinoids. The customizable nature of supercritical fluids allows for high extract selectivity and purity. Because high-pressure systems are required, there is a significant initial equipment investment. Operating expenses can be considerable due to energy consumption and maintenance needs. The ability to reuse CO_2 and high-quality extracts can justify expenses in high-value applications [44].

4.6.2.4 *Plasma-Assisted Extraction (PAE)*

PAE uses plasma to increase extraction by producing reactive species that aid in the breakdown of cell walls and the release of chemicals. PAE is promising for the extraction of heat-sensitive and hydrophobic compounds, although it is still in development with ongoing studies to improve its efficiency and applicability. Equipment costs range from moderate to expensive, depending on the plasma generating system. Operating expenses are often inexpensive due to the use of few solvents and the possibility of continuous processing. The method's innovation may result in greater initial expenditures, but it has the potential to minimize long-term expenses [45].

4.7 Integration of Multiple Extraction Techniques

4.7.1 Hybrid Extraction Methods

Hybrid extraction methods combine two or more extraction techniques to improve bioactive compound recovery rates. These methods seek to maximize the benefits of each technique while minimizing its drawbacks. Microwave-ultrasound–assisted extraction (MUAE) combines microwave heating and cavitation effects. This synergy can result in faster and more efficient extraction. Microwave-supercritical fluid extraction combines microwave heating with supercritical fluids for selective solvent extraction. This hybrid method can greatly improve the extraction of thermolabile and non-polar compounds. Ultrasound-supercritical fluid extraction (USFE) combines ultrasound's mechanical effects to disrupt cell walls with supercritical fluids' effective solvation properties. This combination is especially effective [46]. USFE combines ultrasound's mechanical impacts to shatter cell walls with supercritical fluids' effective solvation properties. This combination works particularly well for extracting bioactive chemicals from difficult or fibrous matrices. Plasma-ultrasound–assisted extraction leverages plasma-generated reactive species and ultrasound-induced mechanical disruption.

4.7.2 Synergistic Effects and Benefits

When techniques are combined, they frequently yield higher yields and faster extraction rates than when they are used alone. For instance, in MUAE, the combined effects of heat and mechanical force can quicken the disruption of cell walls and the diffusion of solutes. Hybrid techniques provide more control over the parameters of the extraction process, which results in a more focused recovery of the targeted compounds [47]. When combined with procedures like UAE or MAE, the non-thermal character of techniques like PAE and the low operating temperatures of SFE help maintain the bioactivity of sensitive chemicals.

4.8 Challenges and Limitations

The implementation of intensification tactics in green engineering may be impeded by several obstacles and constraints. Economic viability is a major challenge since technological complexity and the requirement for specialized equipment and experience can result in high initial costs. Furthermore, transferring laboratory-optimized techniques to industrial settings may result in inefficiencies and even failures. Region-specific regulatory obstacles make implementation much more difficult. Additionally, possibilities for green engineering approaches may be limited because to the scarcity of sustainable resources. Another difficulty is integrating new technologies into old systems, which frequently causes disruptions to ongoing business. Adoption and advancement of green engineering ideas may be slowed by knowledge gaps. New, enhanced procedures usually have a slow rate of market adoption, which hinders their penetration. Additionally, certain tactics could unintentionally result in additional environmental problems, like higher generation of waste or energy use and maintaining consistent performance through sophisticated monitoring and control systems can be difficult and calls for ongoing supervision.

4.8.1 Technical Challenges

Complex integration of processes for maximum resource and energy efficiency is one of the main technical challenges in green engineering [48]. The requirement to smoothly synchronize various technologies and operations gives birth to this complexity. Compatibility problems and operational requirement variations among systems can hinder integration efforts, resulting in performance or even system breakdowns. Scaling from laboratory to industry, from laboratory-scale prototypes to commercially viable green technologies, offers considerable obstacles when applied to industrial applications on a broad basis [49]. Many times, the original designs must be modified in addition to undergoing rigorous testing and validation in order to retain process efficiency, dependability, and cost-effectiveness at industrial sizes. In order to guarantee that sustainability goals are reached, scaling up also brings with it considerations for energy usage, raw material availability, and waste management, all of which need to be properly controlled. Availability and variability of resources like recycled materials, renewable resources, and sustainable substitutes for traditional inputs are all crucial to green engineering. Nonetheless, there may be difficulties due to the fluctuations in the caliber and accessibility of these resources. In order to sustain dependable operations and satisfy production expectations, it becomes imperative to guarantee stable performance, continuous supply chains, and quality assurance [50].

Technological readiness of green engineering techniques, including sophisticated filtering methods, bio-based materials, and renewable energy source: Technologies are still developing and reaching different degrees of maturity. To encourage wider usage across industries, issues with these technologies' scalability, resilience, and dependability must be resolved [4]. To increase the technological preparedness and competitiveness of green engineering solutions, research and development spending is crucial.

4.8.2 Economic and Regulatory Issues

High initial costs for adopting green engineering solutions frequently necessitate large upfront expenditures for infrastructure upgrades, process redesign, and technology adoption. Businesses may find it difficult to make the initial financial adjustments necessary to switch from conventional to sustainable practices, particularly small- and medium-sized organizations with tight budgets [51]. To promote broad adoption, affordable solutions and financial incentives are required. Market acceptance and competition may be difficult for green technologies and goods to adopt in the market despite perceived higher prices in comparison to traditional alternatives [51]. Market adoption may be hampered by consumers' willingness to pay more for sustainability as well as competitive pricing pressures from traditional items. To get past these obstacles, strategic marketing, product differentiation, and consumer education are essential. Regulatory compliance adhering to strict environmental norms and laws raises additional challenges for green engineering projects [52]. Companies have to invest in monitoring and reporting systems, make sure that environmental regulations are followed, and negotiate complicated regulatory environments. Adaptations to enforcement procedures or regulatory frameworks may have an effect on long-term planning and operational strategies for sustainability.

4.9 Potential Solutions and Future Directions

Advanced Process Optimization: Resource consumption can be optimized by utilizing sophisticated modeling and simulation approaches, such as computational fluid dynamics and process integration tools that improve the general effectiveness of the procedure. Performance metrics can be enhanced, and technical difficulties can be reduced by continuously improving operational procedures and process design [53]. Technological innovation promotes technological innovation in green engineering, and research and development expenditures must be sustained [54]. Novel material development, advances in renewable energy technologies, and the integration of smart systems for real-time monitoring and control are some of the areas of focus. Initiatives for cooperative research involving government, business, and academia can hasten the transfer of technology and innovation. Policy support and incentives through encouraging laws, grants, and other financial aid, governments, and regulatory agencies significantly contribute to the advancement of green engineering. Policies that encourage industry adoption and create a favorable environment include carbon pricing schemes, tax incentives for green investments, and awards for sustainable innovation setting for environmentally friendly actions [55].

For circular economy practices by adopting circular economy ideas like resource recovery, waste reduction, and product lifespan extension, industrial operations can become much more sustainable. Strategies that encourage resource efficiency and lessen environmental effect include developing eco-industrial parks, encouraging closed-loop systems and designing items for recycling and disassembly. In international cooperation and standards for green [66] engineering methods to be standardized worldwide, international cooperation and standardization are necessary [56]. Encouraging knowledge sharing, exchanging best practices, and creating uniform standards for sustainability measures can all help to accelerate the global adoption of green techniques and technologies.

4.9.1 Recent Innovations in Green Extraction Techniques

The need for sustainability in all global enterprises is causing a rapid revolution in the field of green extraction processes. Green extraction seeks to replace traditional solvent–based techniques with eco-friendly substitutes that use less energy, produce fewer waste products, and improve the quality of the extracted goods. The emergence of new technology, continuous research and development initiatives, changing market patterns, and bright future possibilities are driving this paradigm shift. In new technologies, Figure 4.5 shows the most cutting-edge techniques in green extraction technologies offer increased productivity, less of an adverse effect on the environment, and better-quality products. SFE is a state-of-the-art method that removes chemicals without leaving behind solvents by using materials like carbon dioxide (CO_2) at particular pressure and temperature conditions. SFE is preferred because it characterizes the capacity to generate high-purity extracts appropriate for use in premium natural products, herbal supplements, and pharmaceuticals. UAE is another noteworthy development [57]. It uses high-frequency sound waves to break down cell structures and make it easier for bioactive substances to be released from plant materials. UAE is highly prized for its quick extraction rates, low solvent consumption, and ability to preserve heat-sensitive chemicals, which makes it perfect for the food, cosmetic, and nutraceutical sectors.

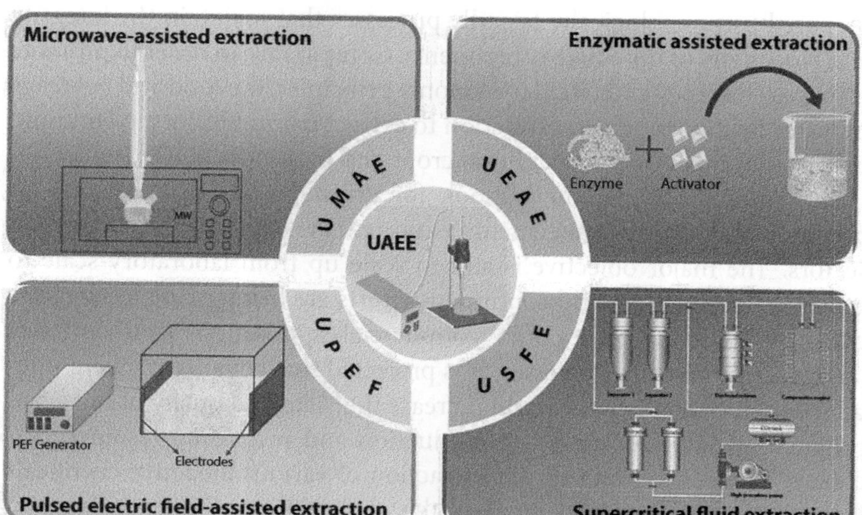

Figure 4.5 Types of ultrasound assisted extraction process.

In research and development, ongoing attempts to improve current technologies and investigate new avenues are essential to the future of green extraction processes. The development of sustainable solvents produced from renewable sources and the optimization of extraction procedures and extraction efficiency depend heavily on interdisciplinary collaboration between chemists, engineers, biologists, and environmental scientists. Because nanotechnology allows for regulated release mechanisms and improves the extraction of complex bioactive compounds from natural sources, it offers great potential for use in green extraction. Furthermore, the investigation of substitute solvents such deep eutectic solvents (DES) and ionic liquids highlights a change toward more environmentally friendly and biologically compatible extraction techniques. These solvents are in line with the ideas of sustainable development and green chemistry because of their adaptable qualities, low toxicity, and biodegradability. For industrial trends, green extraction methods are becoming increasingly popular in a variety of industries, such as agriculture, food and beverage, cosmetics, and medicines. The necessity for sustainable production methods, customer desire for natural and eco-friendly products, and regulatory pressures are all driving this change. With the aim of satisfying strict regulations, guaranteeing product safety, and improving their competitiveness in the market, manufacturers are putting more and more money into green extraction technologies. For example, there is a rising trend in the pharmaceutical and nutraceutical industries to use botanical extracts that are prepared using ecologically friendly processes that maintain the integrity and bioactivity of the active components. Comparably, in order to promote resource efficiency and circular economy principles, the food and beverage industry is utilizing green extraction to extract natural flavors, colors, and useful substances from agricultural crops and by-products [58].

In future prospects, the outlook for green extraction techniques is bright, with major advancements expected in a number of important sectors. The major objective is still to scale up from laboratory-scale to industrial production, and continuous efforts are being made to integrate cutting-edge technologies into commercial applications and optimize large-scale extraction processes. It is projected that advancements in integrated extraction systems would increase the yield and purity of extracted products while reducing waste production and increasing resource efficiency. In order to improve the extraction of various bioactive chemicals from natural sources and increase the applicability of green extraction techniques across many industries, multi-stage extraction technologies are also being investigated.

Collaboration and knowledge sharing is about overcoming current obstacles and realizing the full potential of green extraction technologies. Cooperation between academic institutions, industry players, and policymakers will be essential [24]. Infrastructure, workforce development, and technology transfer projects must all be funded in order to support the global adoption and commercialization of sustainable extraction techniques. In addition, multinational partnerships and knowledge-sharing networks can help resolve regulatory harmonization concerns, foster best practices, and spur innovation—all of which could be obstacles to the wider implementation of green extraction methods [65].

4.9.2 Case Studies and Real-World Applications

In green engineering, intensification techniques are a critical means of attaining efficiency and sustainability in a range of industrial operations. These tactics include a variety of cutting-edge tools and techniques meant to minimize waste production, maximize energy efficiency, and cut down on resource usage [59]. The effectiveness of intensification tactics is demonstrated *via* case studies and practical applications in a variety of industries, including manufacturing and agriculture, underscoring their revolutionary influence on both economic viability and environmental stewardship. The chemical sector offers a compelling case study for intensification techniques, as processes that were formerly marked by high energy consumption and waste generation are undergoing revolutionary change [60]. For example, the use of PI techniques like catalytic processes and continuous flow reactors has greatly increased efficiency while decreasing the environmental impact. Businesses have disclosed significant cuts in the amount of trash produced and the number of raw materials used, resulting in financial savings and environmental advantages. The development of improved solar technology represents a significant application of intensification tactics in the field of renewable energy.

Researchers and engineers have been able to lower the environmental impact of solar panel production and improve energy conversion efficiency through advancements in photovoltaic cell design and manufacturing techniques. These developments help solar energy become more widely used as a sustainable replacement for fossil fuels by maximizing the use of resources and simplifying the manufacturing process. Intensification techniques play a key role in the development of infrastructure and urban planning, helping to create more sustainable cities. Green building technologies [68] allow for the construction of ecologically friendly structures with lower energy consumption and greenhouse gas emissions. Examples of these technologies include renewable energy integration, passive solar design, and effective insulation. These structures act as role models for sustainable development, demonstrating how intensification techniques [67] can be used to lessen urbanization's environmental impact. The implementation of intensification measures finds a strong home in the transportation industry [63].

Automobile electrification and recent developments in battery technology, which provide greener and more energy-efficient substitutes for traditional internal combustion engines, are revolutionizing the automotive sector [61]. Intensification measures help to mitigate the environmental impact of transportation networks and reduce air pollution by encouraging the adoption of electric and hybrid vehicles. The use of intensification techniques in green engineering solves environmental issues and promotes innovation and economic growth in a variety of applications [64]. Businesses and sectors adopting these tactics benefit from lower operating costs, better product quality, and increased resource efficiency, which gives them a competitive edge. By offering incentives for the adoption of green technologies, sponsoring research and development projects, and creating legislative frameworks that promote sustainable behaviors, governments and legislators play a critical role in assisting these efforts. In the future, the ongoing global sustainability targets may be possible if intensification strategies are widely adopted and continue to progress. Through the judicious use of investment [69], interdisciplinary collaboration, and technological innovation, stakeholders may quicken the shift toward a future that is more resilient and sustainable. The revolutionary potential of intensification tactics in green engineering is compellingly demonstrated by case studies and real-world applications, which encourage further innovation and pave the way for a more affluent and environmentally conscious world [62].

4.10 Conclusion

In conclusion, our investigation into PI in the field of green engineering emphasizes how crucial it is to forming a sustainable future. PI is the embodiment of effectiveness, resource optimization, and environmental care, as demonstrated by numerous research and examples. PI is essential in many different industries because it reduces waste, energy use, and emissions while increasing production efficiency and product quality. In the long run, PI will depend on incorporating cutting-edge technology like improved materials, machine learning, and artificial intelligence. Along with improved optimization, these technologies also hold the potential to create whole new, economically and environmentally sustainable processes. But accomplishing these objectives will call for coordinated efforts by academics, business leaders, and legislators. Researchers need to keep pushing the borders of knowledge, investigating cross-disciplinary partnerships that fill in gaps in the fields of environmental science, materials science, engineering, and chemistry. Industry professionals must simultaneously prioritize funding for environmentally friendly technology and procedures, using their knowledge to put into practice scalable solutions that maintain environmental standards without sacrificing output. In addition, legislative frameworks need to change in order to encourage and promote these developments and create an atmosphere that is favorable to innovation and uptake. Stakeholders can collectively steer the course of green engineering toward a future where PI forms the basis of sustainable development and ensures a harmonious balance between industrial advancement and environmental preservation for future generations by adopting these principles and taking proactive measures.

AI disclosure statement: The author acknowledges the use of AI-based tools for minor language editing or grammatical correction. However, all intellectual contributions, analyses, and conclusions presented in this chapter are entirely the author's own.

References

1. Balaraman, H., Selvasembian, R., Rangarajan, V., Rathnasamy, S., Sustainable and green engineering insights on deep eutectic solvents toward the extraction of nutraceuticals. *ACS Sustain. Chem. Eng.*, *9*, 34, 11290–11313, 2021.
2. Bradu, P., Biswas, A., Nair, C., Sreevalsakumar, S., Patil, M., Kannampuzha, S., Gopalakrishnan, A.V., Recent advances in green technology and Industrial Revolution 4.0 for a sustainable future. *Environ. Sci. Pollut. Res.*, *30*, 60, 124488–124519, 2023.
3. Klemeš, J.J., Varbanov, P.S., Alwi, S.R.W., Manan, Z.A., Van Fan, Y., Chin, H.H., *Sustainable process integration and intensification: Saving energy, water and resources*, Walter de Gruyter GmbH & Co. KG, UTM, Johor, Malaysia, 2023.
4. Sharma, S.K. (Ed.), *Green chemistry: Its role in achieving sustainable development goals*, vol. 1, CRC Press, Boca Raton, FL, 2023.
5. Wasewar, K.L., Singh, S., Kansal, S.K., Process intensification of treatment of inorganic water pollutants, in: *Inorganic pollutants in water*, Elsevier, Nagpur & Chandigarh, India, pp. 245–271, 2020.

6. Constantino, D.S., Dias, M.M., Silva, A.M., Faria, J.L., Silva, C.G., Intensification strategies for improving the performance of photocatalytic processes: A review. *J. Clean. Prod.*, *340*, 130800, 2022.
7. Zhu, J., Hao, L., Wei, H., Sustainable concept design including economic, environment and inherent safety criteria: Process intensification-reactive pressure swing distillation. *J. Clean. Prod.*, *314*, 127852, 2021.
8. Adamu, A., Russo-Abegão, F., Boodhoo, K., Process intensification technologies for CO2 capture and conversion–a review. *BMC Chem. Eng.*, *2*, 1, 2, 2020.
9. Boodhoo, K.V.K., Flickinger, M.C., Woodley, J.M., Emanuelsson, E.A.C., Bioprocess intensification: A route to efficient and sustainable biocatalytic transformations for the future. *Chem. Eng. Process. Process Intensif.*, *172*, 108793, 2022. Buckingham, J., Reina, T.R., Duyar, M.S., Recent advances in carbon dioxide capture for process intensification. *Carbon Capture Sci. Technol.*, *2*, 100031, 2022.
10. De Risi, C., Bortolini, O., Brandolese, A., Di Carmine, G., Ragno, D., Massi, A., Recent advances in continuous-flow organocatalysis for process intensification. *React. Chem. Eng.*, *5*, 6, 1017–1052, 2020.
11. Hakke, V., Sonawane, S., Anandan, S., Sonawane, S., Ashokkumar, M., Process intensification approach using microreactors for synthesizing nanomaterials— A critical review. *Nanomaterials*, *11*, 1, 98, 2021.
12. Kong, Z.Y., Sánchez-Ramírez, E., Yang, A., Shen, W., Segovia-Hernández, J.G., Sunarso, J., Process intensification from conventional to advanced distillations: Past, present, and future. *Chem. Eng. Res. Des.*, *188*, 378–392, 2022.
13. López-Guajardo, E.A., Delgado-Licona, F., Álvarez, A.J., Nigam, K.D., Montesinos-Castellanos, A., Morales-Menendez, R., Process intensification 4.0: A new approach for attaining new, sustainable and circular processes enabled by machine learning. *Chem. Eng. Process.-Process Intensif.*, *180*, 108671, 2022.
14. Rivas, D.F., Boffito, D.C., Faria-Albanese, J., Glassey, J., Afraz, N., Akse, H., Weber, R.S., Process intensification education contributes to sustainable development goals. Part 1. *Educ. Chem. Eng.*, *32*, 1–14, 2020.
15. Rao, M.V., Sengar, A.S., Sunil, C.K., Rawson, A., Ultrasonication-A green technology extraction technique for spices: A review. *Trends Food Sci. Technol.*, 116, 975–991, 2021.
16. Gouda, M., Bekhit, A.E.D., Tang, Y., Huang, Y., Huang, L., He, Y., Li, X., Recent innovations of ultrasound green technology in herbal phytochemistry: A review. *Ultrason. Sonochem.*, 73, 105538, 2021.
17. Wang, J., Jing, W., Tian, H., Liu, M., Yan, H., Bi, W., Chen, D.D.Y., Investigation of deep eutectic solvent-based microwave-assisted extraction and efficient recovery of natural products. *ACS Sustain. Chem. Eng.*, 8, 32, 12080–12088, 2020.
18. Wen, L., Zhang, Z., Zhao, M., Senthamaraikannan, R., Padamati, R.B., Sun, D.W., Tiwari, B.K., Green extraction of soluble dietary fibre from coffee

silverskin: impact of ultrasound/microwave-assisted extraction. *Int. J. Food Sci. Technol.*, 55, 5, 2242–2250, 2020.

19. Alvi, T., Asif, Z., Khan, M.K.I., Clean label extraction of bioactive compoundsfrom food waste through microwave-assisted extraction technique-A review. *Food Biosci.*, *46*, 101580, 2022.
20. Nour, A.H., Oluwaseun, A.R., Nour, A.H., Omer, M.S., Ahmed, N., Microwave-assisted extraction of bioactive compounds, in: *Microwave heating. Electromagnetic fields causing thermal and non-thermal effects*, pp. 1–31, 2021.
21. da Rocha, C.B. and Noreña, C.P.Z., Microwave-assisted extraction and ultrasound-assisted extraction of bioactive compounds from grape pomace. *Int. J. Food Eng.*, *16*, 1-2, 20190191, 2020.
22. Sharma, B.R., Kumar, V., Kumar, S., Panesar, P.S., Microwave assisted extraction of phytochemicals from Ficus racemosa. *Curr. Res. Green Sustain. Chem.*, *3*, 100020, 2020.
23. Mushtaq, A., Roobab, U., Denoya, G.I., Inam-Ur-Raheem, M., Gullón, B., Lorenzo, J.M., Aadil, R.M., Advances in green processing of seed oils using ultrasound-assisted extraction: A review. *J. Food Process. Preserv.*, *44*, 10, e14740, 2020.
24. Shen, L., Pang, S., Zhong, M., Sun, Y., Qayum, A., Liu, Y., Ren, X., A comprehensive review of ultrasonic assisted extraction (UAE) for bioactive components: Principles, advantages, equipment, and combined technologies. *Ultrason. Sonochem.*, 91, 106646, 2023.
25. Rashid, R., Wani, S.M., Manzoor, S., Masoodi, F.A., Dar, M.M., Green extraction of bioactive compounds from apple pomace by ultrasound assisted natural deep eutectic solvent extraction: Optimisation, comparison and bioactivity. *Food Chem.*, *398*, 133871, 2023.
26. Yusoff, I.M., Taher, Z.M., Rahmat, Z., Chua, L.S., A review of ultrasound-assisted extraction for plant bioactive compounds: Phenolics, flavonoids, thymols, saponins and proteins. *Food Res. Int.*, *157*, 111268, 2022.
27. Lameirão, F., Pinto, D., Vieira, E.F., Peixoto, A.F., Freire, C., Sut, S., Rodrigues, F., Green-sustainable recovery of phenolic and antioxidant compounds from industrial chestnut shells using ultrasound-assisted extraction: Optimization and evaluation of biological activities *in vitro*. *Antioxidants*, *9*, 3, 267, 2020.
28. Sapkale, G.N., Patil, S.M., Surwase, U.S., Bhatbhage, P.K., Supercritical fluid extraction. *Int. J. Chem. Sci.*, *8*, 2, 729–743, 2010.
29. Clifford, A. A., Williams, J. R., Introduction to supercritical fluids and their applications, in: *Supercritical fluid technology*, Humana Press, Totowa, New Jersey, pp. 1–16, 2000.
30. Clifford, A.A., Williams, J. R., Introduction to supercritical fluids and their applications, in: *Supercritical fluid technology*, Humana Press, Totowa, New Jersey, pp. 1–16, 2000.
31. Shinde, V. and Mahadik, K., Supercritical fluid extraction: A new technology to herbals. *Int. J. Herb. Med.*, *7*, 27–34, 2019.

32. Lee, W.J., Suleiman, N., Hadzir, N.H.N., Chong, G.H., Supercritical fluids for the extraction of oleoresins and plant phenolics, in: *Green Sustainable Process for Chemical and Environmental Engineering and Science*, Elsevier, Selangor, Malaysia, pp. 279–328, 2020.
33. Da Silva, R.P., Rocha-Santos, T.A., Duarte, A.C., Supercritical fluid extraction of bioactive compounds. *TrAC, Trends Anal. Chem.*, *76*, 40–51, 2016.
34. Boenig, H., *Plasma science and technology*, Cornell University Press, Ithaca, New York, 2019.
35. Conrads, H. and Schmidt, M., Plasma generation and plasma sources. *Plasma Sources Sci. Technol.*, *9*, 4, 441, 2000.
36. Heydari, M., Carbone, K., Gervasi, F., Parandi, E., Rouhi, M., Rostami, O., Mohammadi, R., Cold plasma-assisted extraction of phytochemicals: a review. *Foods*, *12*, 17, 3181, 2023.
37. Chen, X., Cheng, Y., Li, T., Cheng, Y., Characteristics and applications of plasma assisted chemical processes and reactors. *Curr. Opin. Chem. Eng.*, *17*, 68–77, 2017.
38. Chan, C.H., Yusoff, R., Ngoh, G.C., Kung, F.W.L., Microwave-assisted extractions of active ingredients from plants. *J. Chromatogr. A*, *1218*, 37, 6213–6225, 2011.
39. Zhang, M., Zhao, J., Dai, X., Li, X., Extraction and analysis of chemical compositions of natural products and plants. *Separations*, *10*, 12, 598, 2023.
40. Reverchon, E. and De Marco, I., Supercritical fluid extraction and fractionation of natural matter. *J. Supercrit. Fluids*, *38*, 2, 146–166, 2006.
41. Alara, O.R., Abdurahman, N.H., Ukaegbu, C.I., Extraction of phenolic compounds: A review. *Curr. Res. Food Sci.*, *4*, 200–214, 2021.
42. Esclapez, M.D., García-Pérez, J.V., Mulet, A., Cárcel, J.A., Ultrasound-assisted extraction of natural products. *Food Eng. Rev.*, *3*, 108–120, 2011.
43. Uwineza, P.A. and Waśkiewicz, A., Recent advances in supercritical fluid extraction of natural bioactive compounds from natural plant materials. *Molecules*, *25*, 17, 3847, 2020.
44. Xi, J., Wang, Y., Zhou, X., Wei, S., Zhang, D., Cold plasma pretreatment technology for enhancing the extraction of bioactive ingredients from plant materials: A review. *Ind. Crops Prod.*, *209*, 117963, 2024.
45. Rifna, E.J., Misra, N.N., Dwivedi, M., Recent advances in extraction technologies for recovery of bioactive compounds derived from fruit and vegetable waste peels: A review. *Crit. Rev. Food Sci. Nutr.*, *63*, 6, 719–752, 2023.
46. Khadhraoui, B., Ummat, V., Tiwari, B.K., Fabiano-Tixier, A.S., Chemat, F., Review of ultrasound combinations with hybrid and innovative techniques for extraction and processing of food and natural products. *Ultrason. Sonochem.*, *76*, 105625, 2021.
47. Nhien, L.C., Long, N.V.D., Kim, S., Lee, M., Techno-economic assessment of hybrid extraction and distillation processes for furfural production from lignocellulosic biomass. *Biotechnol. Biofuels*, *10*, 1–12, 2017.

48. Zhang, Z.S., Xie, Q.F., Che, L., Synergistic effects of ultrasound and extraction solvent on the bioactive compound in kenaf seed oil. *J. Food Sci. Technol.*, *57*, 2118–2128, 2020.
49. Guo, Z., Sun, Y., Pan, S.Y., Chiang, P.C., Integration of green energy and advanced energy-efficient technologies for municipal wastewater treatment plants. *Int. J. Environ. Res. Public Health*, *16*, 7, 1282, 2019.
50. Baptista, J.M., Sagu, J.S., KG, U.W., Lobato, K., State-of-the-art materials for high power and high energy supercapacitors: Performance metrics and obstacles for the transition from lab to industrial scale–A critical approach. *Chem. Eng. J.*, *374*, 1153–1179, 2019.
51. Fraga-Lamas, P. and Fernández-Caramés, T.M., A review on blockchain technologies for an advanced and cyber-resilient automotive industry. *IEEE Access*, *7*, 17578–17598, 2019.
52. Karadag, H., *Strategic financial management for small and medium sized companies*, Emerald Group Publishing, United Kingdom, 2015.
53. Kowalska-Pyzalska, A., What makes consumers adopt to innovative energy services in the energy market? A review of incentives and barriers. *Renew. Sustain. Energy Rev.*, *82*, 3570–3581, 2018.
54. Allen, D.T. and Shonnard, D.R., *Green engineering: Environmentally conscious design of chemical processes*, Pearson Education, New York, 2001.
55. Ferdows, K. and De Meyer, A., Lasting improvements in manufacturing performance: in search of a new theory. *J. Oper. Manage.*, *9*, 2, 168–184, 1990.
56. Alnour, M., Önden, A., Hasseb, M., Önden, I., Rehman, M.Z., Esquivias, M.A., Hossain, M.E., The Asymmetric Role of Financial Commitments to Renewable Energy Projects, Public R&D Expenditure, and Energy Patents in Sustainable Development Pathways. *Sustainability*, *16*, 13, 5503, 2024.
57. Owen, R., Brennan, G., Lyon, F., Enabling investment for the transition to a low carbon economy: Government policy to finance early stage green innovation. *Curr. Opin. Environ. Sustain.*, *31*, 137–145, 2018.
58. Krause, J., Through harmonization of national technical regulations to more sustainability in engineering business: How to stay compliant, efficient, and sustainable in international engineering business, in: *Imagination, creativity, and responsible management in the fourth industrial revolution*, IGI Global, Germany, pp. 263–288, 2019.
59. Herrero, M. and Ibañez, E., Green extraction processes, biorefineries and sustainability: Recovery of high added-value products from natural sources. *J. Supercrit. Fluids*, *134*, 252–259, 2018.
60. Belwal, T., Ezzat, S.M., Rastrelli, L., Bhatt, I.D., Daglia, M., Baldi, A., Atanasov, A.G., A critical analysis of extraction techniques used for botanicals: Trends,priorities, industrial uses and optimization strategies. *TrAC, Trends Anal. Chem.*, *100*, 82–102, 2018.
61. Yoon, H.S., Kim, E.S., Kim, M.S., Lee, J.Y., Lee, G.B., Ahn, S.H., Towards greener machine tools–A review on energy saving strategies and technologies. *Renew. Sustain. Energy Rev.*, *48*, 870–891, 2015.

62. Boodhoo, K. and Harvey, A. (Eds.), *Process intensification technologies for green chemistry: engineering solutions for sustainable chemical processing*, John Wiley & Sons, 2013.
63. Pretty, J. and Bharucha, Z.P., Sustainable intensification in agricultural systems. *Ann. Bot.*, *114*, 8, 1571–1596, 2014.
64. Rivas, D.F. and Cintas, P., On an intensification factor for green chemistry and engineering: The value of an operationally simple decision-making tool in process assessment. *Sustain. Chem. Pharm.*, *27*, 100651, 2022.
65. Stankiewicz, A. and Moulijn, J.A., Process intensification. *Ind. Eng. Chem. Res.*, *41*, 8, 1920–1924, 2002.
66. Abraham, M.A. and Nguyen, N., Green engineering: Defining the principles results from the sandestin conference. *Environ. Prog.*, *22*, 4, 233–236, 2003.
67. García-Serna, J., Pérez-Barrigón, L., Cocero, M.J., New trends for design towards sustainability in chemical engineering: Green engineering. *Chem. Eng. J.*, *133*, 1-3, 7–30, 2007.
68. Sarte, S.B., *Sustainable infrastructure: The guide to green engineering and design*, John Wiley & Sons, Hoboken, New Jersey, 2010.
69. Shonnard, D.R., Kicherer, A., Saling, P., Industrial applications using BASF eco-efficiency analysis: perspectives on green engineering principles. *Environ. Sci. Technol.*, *37*, 23, 5340–5348, 2003.

5

Green Nanoemulsions for Sustainable Applications

Ashwini Padole[1], Sreelakshmi K.P.[2], Utpal Roy[2] and Vivek Rangarajan[1]*

[1]*Department of Chemical Engineering, Birla Institute of Technology and Science Pilani KK Birla Goa Campus, Sancoale, India*

[2]*Department of Biological Sciences, Birla Institute of Technology and Science Pilani K.K. Birla Goa Campus, Goa, India*

Abstract

The chapter explores the role of green nanoemulsions, a versatile class of colloidal dispersions with droplet sizes ranging from 20 to 200 nm, in various industries such as pharmaceuticals, cosmetics, and food. It highlights the advantages of using sustainable, biodegradable ingredients in their formulation, aligning with green chemistry principles. These eco-friendly nanoemulsions minimize environmental impacts by avoiding hazardous chemicals and utilizing renewable resources such as plant-based oils and natural surfactants. The chapter discusses different preparation methods, including high-energy and low-energy emulsification techniques, as well as their benefits in improving the stability and bioavailability of active compounds. Additionally, the potential applications of green nanoemulsions in enhancing drug delivery, food preservation, and cosmetics are elaborated upon, offering promising solutions for eco-conscious industries. The future of nanoemulsions is closely tied to continued advancements in nanotechnology and sustainability, ensuring broader industrial applications.

Keywords: Green nanoemulsions, sustainable nanoemulsion formulations, green chemistry principles, oil-in-water (O/W) emulsions, water-in-oil (W/O) emulsions, bi-continuous nanoemulsions

**Corresponding author*: vivekr@goa.bits-pilani.ac.in

Senthilkumar Rathnasamy, Vivek Rangarajan and Umile Gianfranco Spizzirri (eds.) Green Engineering and Sustainable Technology: Tools and Techniques, (135–160) © 2026 Scrivener Publishing LLC

5.1 Introduction

Nanoemulsions have garnered significant attention from researchers and industries over the past decades due to their versatility across various applications. Defined by their minuscule droplet sizes and distinctive properties, they represent a key area of interest in pharmaceuticals, food science, drug delivery, cosmetics, and material synthesis (Figure 5.1). Nanoemulsions are colloidal dispersions with droplet sizes ranging from 20 to 200 nm and can exist as either oil-in-water (O/W) or water-in-oil (W/O) systems [1]. Typically composed of oil, surfactants, and water, these emulsions exhibit enhanced kinetic stability, which has made them particularly attractive in the cosmetics industry. Their ability to improve the delivery of active ingredients, enhance product texture, and increase skin absorption makes them ideal for use in formulations such as moisturizers, sunscreens, and anti-aging treatments [2, 3].

Recent scientific advancements have opened new avenues in nanoemulsion research, particularly in the development of green nanoemulsions, while also driving progress in existing fields of its applications. These green nanoemulsions align with the fundamental principles of green chemistry, such as prevention, atom economy, and design for degradation [1]. By avoiding hazardous substances and prioritizing the use of renewable, eco-friendly materials, they contribute to reducing pollution and improving product safety [4, 5]. This chapter explores the rationale behind adopting

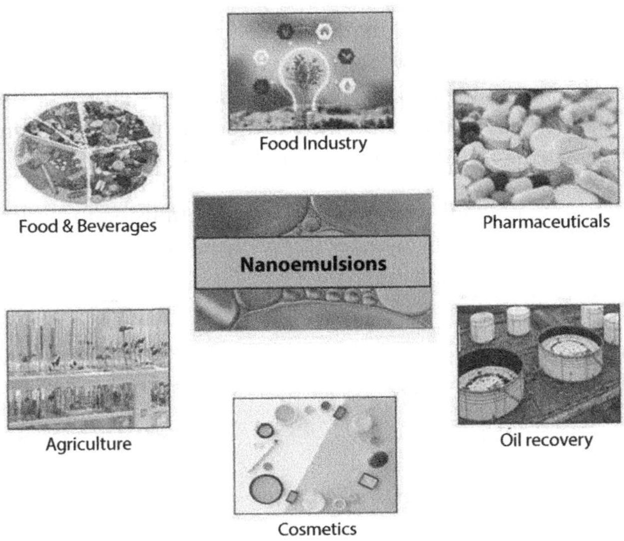

Figure 5.1 Applications of nanoemulsions.

green nanoemulsions, the sustainable ingredients used in their formulation, their applications across various sectors, the role of base and essential oils in cosmetic products, and the integration of green energy into their production processes, all supported by recent research findings.

5.2 Fundamentals of Nanoemulsions

5.2.1 Composition and Types

Nanoemulsions are distinguished by their nanoscale droplet size and superior stability compared to conventional emulsions. Their small droplet size offers several advantages, including improved stability, enhanced bioavailability, and more efficient encapsulation of active ingredients, making them highly effective in a wide range of applications [6]. Based on the nature of the continuous and dispersed phases, nanoemulsions can be categorized into three types are us follows.

5.2.1.1 W/O Nanoemulsions

In W/O emulsions, water droplets are dispersed within a continuous oil phase. This structure is particularly beneficial in applications requiring a lipophilic environment, such as transdermal drug delivery systems and certain cosmetic formulations [7]. The small droplet size enhances the stability of the emulsion and allows for better penetration of active ingredients into the skin, making W/O nanoemulsions highly valuable for both pharmaceutical and cosmetic applications [8].

5.2.1.2 O/W Nanoemulsions

O/W nanoemulsions, in which oil droplets are dispersed within a continuous aqueous phase, are widely used in various industries such as food and beverages, pharmaceuticals (particularly in drug delivery), and cosmetics. Their hydrophilic environment and compatibility with water-based systems make them ideal for these applications [7].

5.2.1.3 Bi-Continuous Nanoemulsions

Bi-continuous nanoemulsions, although less common, offer unique advantages in applications that require the simultaneous presence of both hydrophilic and lipophilic environments. This structure allows for enhanced versatility, making them valuable in specialized formulations where the interaction of water and oil phases is essential [9].

5.2.1.4 *Size and Stability*

This type of emulsion, often referred to as micro-sized emulsions, nanoscale emulsions, or ultrafine droplet emulsions, is gaining significant attention from researchers due to its unique properties and versatility across various applications [10]. Potential uses include selective drug targeting, the food processing industry, and, notably, skincare products [9]. The small droplet size of nanoemulsions enhances their stability, preventing gravitational separation and ensuring a more consistent formulation.

5.2.1.5 *Optical Properties*

Nanoemulsions can appear either clear or translucent, depending on the droplet size and the refractive index difference between the dispersed and continuous phases. The nanoscale droplet size enables efficient light scattering, which contributes to the optical clarity of nanoemulsions [11, 12].

5.3 Benefits, Drawbacks and Key Attributes of Nanoemulsions

The advantages and disadvantages explaining why nanoemulsion is appealing in the personal care, cosmetics industry, and healthcare are shown in Figure 5.2 [13].

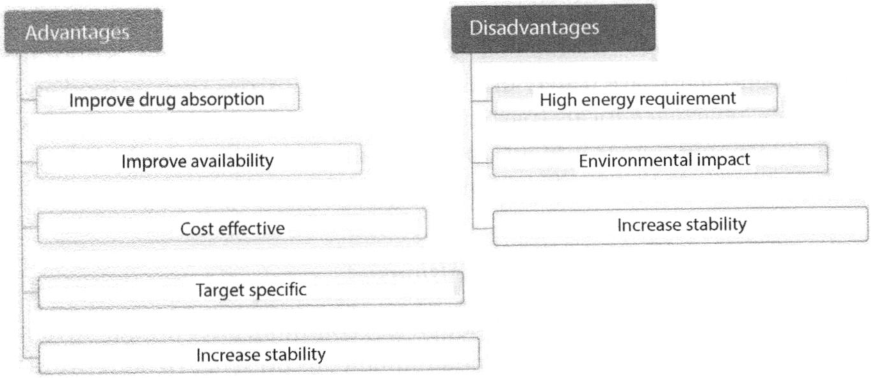

Figure 5.2 Advantages and disadvantages of nanoemulsion [13].

5.4 Challenges and Limitations of Nanoemulsions

Although nanoemulsions offer significant advantages as a delivery system, they face several challenges and limitations in their preparation, characterization, and application [11, 12]. Figure 5.3 shows the major challenges associated with nanoemulsions, including safety concerns, component selection, and preparation methods [13].

5.4.1 High Cost of Formulation

The production of nanoemulsions is expensive due to the difficulty in reducing droplet size, which requires specialized equipment and techniques. For example, the homogenizers used in nanoemulsion production are costly to operate, and methods like micro-fluidization and ultrasonication involve substantial financial investment.

5.4.2 Environmental Sensitivity

Nanoemulsions are sensitive to ecological factors such as temperature and pH, which can affect their stability.

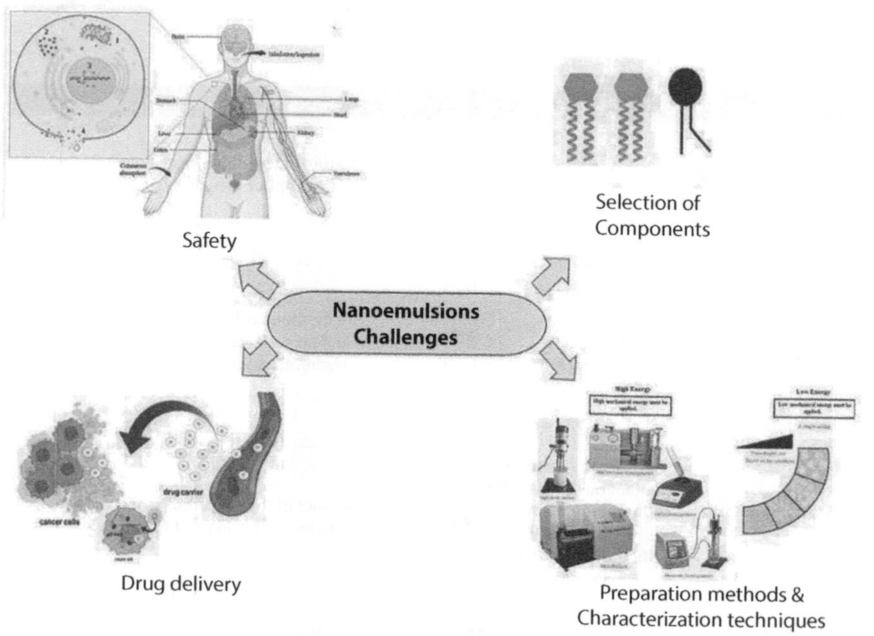

Figure 5.3 Challenges of nanoemulsions [13].

5.4.3 Limited Research

There is still insufficient research on the production of submicron droplets and the specific roles of surfactants and co-surfactants in the nano emulsification process.

5.4.4 Understanding of Interfacial Chemistry

The interfacial chemistry involved in nanoemulsion production is not yet fully understood, presenting a knowledge gap that affects formulation optimization.

5.5 Synthesis and Characterization of Nanoemulsions

5.5.1 Selection of Components

Below the flowchart, Figure 5.4 shows how nanoemulsion components are selected. Firstly, the type of nanoemulsion is decided (i.e., oil/water, water/oil, or multiple). Next, the components of the nanoemulsion are decided. Nanoemulsions are prepared using oils, surfactants, co-surfactants, and aqueous phase [15, 16].

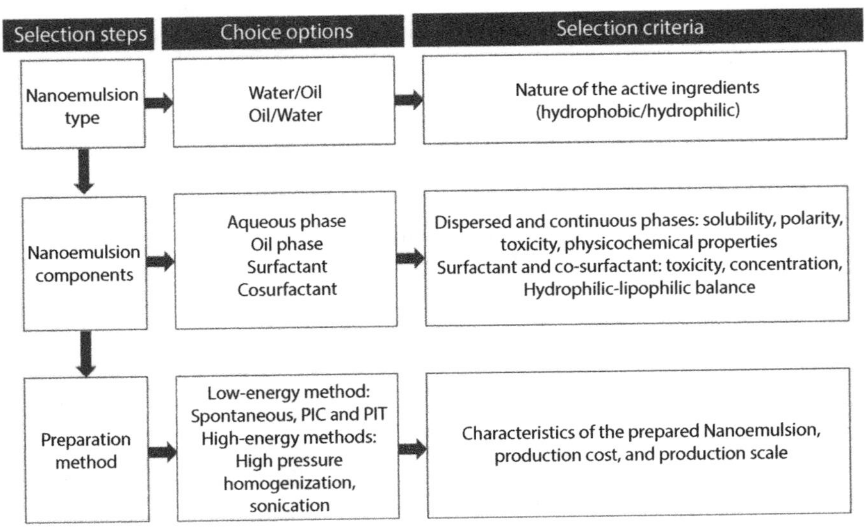

Figure 5.4 Flowchart of selecting the nanoemulsion components [14].

Oil Phase: The oil phase can consist of natural oils (e.g., essential oils and vegetable oils) or synthetic oils. The choice of oil affects the solubility of active ingredients and the stability of the nanoemulsion.
Surfactants: Types: Non-ionic, cationic, anionic, and zwitterionic. Surfactants are Gelucire 44/14, Capryol 90, Cremophor RH 40, Tween 20, Tween 80, etc. Also, non-ionic surfactants are preferred due to lower toxicity and better *in vivo* stability. The selection of surfactant was based on hydrophile-lipophile balance (HLB); for O/W nanoemulsions, an HLB value greater than 10 is needed.
Co-Surfactants: Co-surfactants are often used in conjunction with surfactants to enhance the stability of the nanoemulsion. Examples include alcohols such as ethanol and propylene glycol, which help improve the overall performance of the emulsion.
Aqueous Phase: Typically, water or a water-based solution. The ionic and pH strength of the aqueous phase can influence stability.
Emulsifiers: Surfactants or a surfactant combination that lower interfacial tension between the water and oil phases. Common emulsifiers include [17, 18] non-ionic surfactants (e.g., Tween and Span), ionic surfactants (e.g., sodium dodecyl sulfate), and natural emulsifiers (e.g., lecithin and gum Arabic).

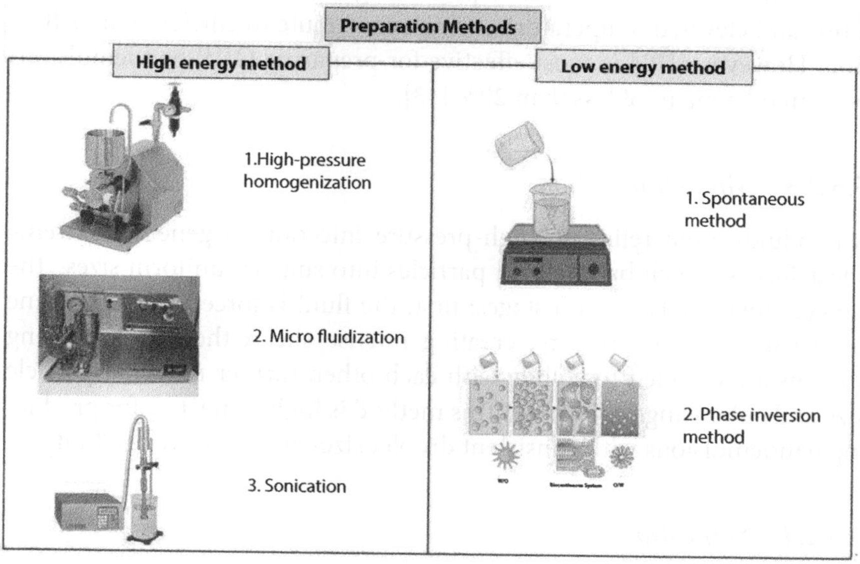

Figure 5.5 Methods of preparation of nanoemulsion [19].

5.5.2 Formulation and Preparation Method of Nanoemulsions

To prepare nanoemulsion, mainly two types of methods are employed: high energy method and low energy method (Figure 5.5).

5.5.2.1 *High-Energy Emulsification Method*

The formulation of nanoemulsions often involves high-energy techniques, such as the use of high-pressure homogenizers and microfluidizers, to break down the oil and water phases into ultra-fine droplets. High-pressure homogenizers are widely employed because the process relies on the application of intense shear forces, reducing droplet sizes to the nanoscale [13, 20]. During emulsification, pressures ranging from 500 to 5,000 psi are applied, forcing the mixture through a narrow aperture. This creates strong hydraulic shear, resulting in an emulsion with extremely small particle sizes, which enhances the stability and effectiveness of the nanoemulsion.

5.5.2.2 *High-Pressure Homogenization (HPH)*

The HPH technique applies pressures up to 350 MPa to induce particle collisions, leading to the formation of nanoemulsion droplets. Compared to low-energy emulsification methods, HPH—when paired with the appropriate surfactants—offers shorter processing times, as well as higher shear stress and elevated temperatures, which contribute to efficient emulsification. However, HPH is most effective for preparing O/W nanoemulsions with an oil content of less than 20% [13].

5.5.2.3 *Microfluidization*

Microfluidization relies on high-pressure injection to generate extreme shear forces, which break down particles into smaller, uniform sizes. The process involves two main stages: first, the fluid is forced through a fine orifice under high pressure, creating intense shear; then, the resulting streams are directed to collide with each other, further reducing particle size and enhancing uniformity. This method is highly effective for producing nanoemulsions with consistent droplet sizes and improved stability.

5.5.2.4 *Sonication*

Sonication is often compared to low-energy techniques, such as spontaneous emulsification, which typically require less costly equipment.

In sonication, ultrasound waves generate mechanical compression and rarefaction, leading to the formation of cavitation bubbles that implode, causing the breakdown of particles. By adjusting the duration or power of the ultrasound treatment, particle size can be reduced to a certain limit; however, beyond this point, further increases in ultrasound intensity or time have no additional effect on particle size reduction [21]. This method is effective for achieving small, uniform droplets with relatively simple equipment.

5.5.2.5 Low-Energy Emulsification Method

Low-energy emulsification relies on techniques like shaking or stirring to form nanoemulsions. While simpler and less costly than high-energy methods, this approach is generally less efficient and tends to produce larger droplet sizes [22]. As a result, low-energy emulsification is often less effective in achieving the fine, stable dispersions typical of nanoemulsions produced through high-energy methods.

5.5.2.6 Spontaneous Methods

Spontaneous emulsification is a simple method for creating nanoemulsions, where dispersion droplets form naturally without the need for high-energy input. However, a significant challenge of this technique is its limitation on the amount of dispersed phase that can be used. To increase the percentage of the dispersed phase, higher levels of surfactants or co-surfactants are required, which can introduce complications such as stability issues, increased toxicity, or undesirable effects on the final product [23].

5.5.2.7 Phase Inversion Method

The phase inversion method is a widely used technique in cosmetics to create stable emulsions. It involves mixing an oil phase with an aqueous phase containing emulsifiers. By adjusting either the temperature or the concentrations of the phases, a phase inversion occurs, where the continuous and dispersed phases switch. This results in a stable emulsion with the desired texture and consistency, making it ideal for formulating products like creams and lotions. This method ensures improved stability and performance in the final product [15]. Figure 5.6 illustrates the different preparation methods of nanoemulsions, comparing high-energy techniques (such as high-pressure homogenization, micro fluidization, and sonication) with low-energy methods (such as spontaneous emulsification and phase inversion).

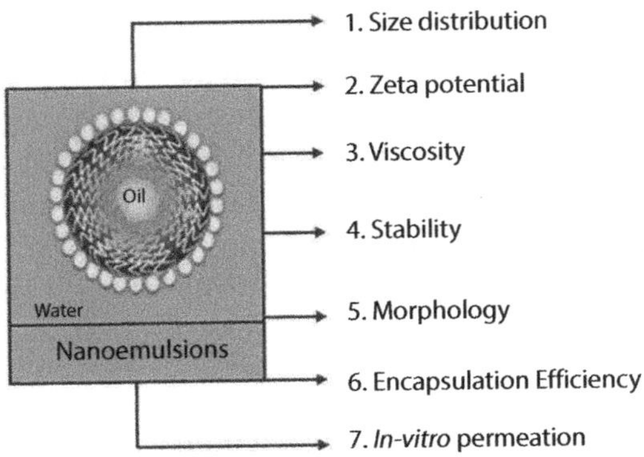

Figure 5.6 A flowchart of characterization techniques for nanoemulsions [19].

5.6 Characterization of Nanoemulsion

A stable nanoemulsion is defined by the absence of internal phase separation, lack of creaming, resistance to microbial degradation, and the retention of key aesthetic qualities such as appearance, color, odor, surface tension, texture, and consistency. These characteristics provide valuable insights into the behavior of nanoemulsions and potential interactions within the formulation, ensuring long-term stability and effectiveness [2]. Comprehensive characterization of nanoemulsions is essential to ensure stability, safety, and effectiveness. Various techniques are employed to analyze particle size, zeta potential, viscosity, stability, and other physical properties. Figure 5.7 summarizes the main characterization techniques used for nanoemulsions.

5.6.1 Particle Size and Distribution

5.6.1.1 Size Distribution

Nanoemulsions should exhibit a narrow size distribution, typically indicated by a low polydispersity index (PDI). A PDI value close to 0 reflects a more uniform size distribution, whereas a PDI above 0.3 suggests a broader distribution, which could lead to instability [24, 25].

5.6.1.2 Measurement Techniques

Particle size and particle size distribution are typically measured using techniques like dynamic light scattering, laser diffraction, or transmission electron microscopy [26–28] or atomic force microscopy, which offers

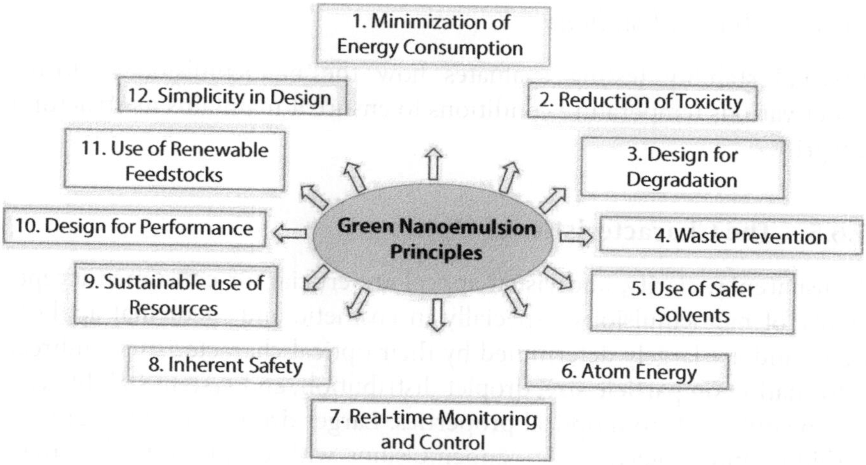

Figure 5.7 Principles of green nanoemulsion [43].

high-resolution imaging of surface topology and particle size. Microscopic analysis provides information about stability, homogeneity, and morphology [29].

5.6.2 Zeta Potential

Electrophoretic light scattering is used to measure the zeta potential, which indicates the surface charge of particles and assesses the stability of nanoemulsions. High absolute zeta potential values (either positive or negative) suggest good stability, as they reduce the likelihood of particle aggregation [30].

5.6.3 Viscosity and Rheology

Rheometers are employed to measure the viscosity and flow behavior of nanoemulsions, both of which are critical for processing and application in various formulations.

5.6.4 Stability Studies

5.6.4.1 Accelerated Stability Testing

This method subjects the nanoemulsion to stress conditions such as temperature variation, centrifugation, and freeze-thaw cycles to evaluate its stability. Centrifugation, in particular, accelerates the separation process, providing a quick assessment of physical stability [30].

5.6.4.2 *Thermal Stability*

Thermal stability testing evaluates how the nanoemulsion performs under various temperature conditions to ensure it maintains its structural integrity.

5.6.5 The Characteristics of Optics

Transparency, clarity, and visual appeal are crucial for consumer acceptability of nanoemulsions, especially in cosmetic and medicinal applications, and are largely determined by their optical characteristics. Indirect information on particle size, droplet distribution, and system stability can also be obtained from optical properties. Larger droplets result in greater turbidity and opacity, while nanoemulsions with droplet sizes less than 100 nm frequently seem transparent or translucent because of less light scattering.

In addition to basic optical measurements, turbidimetric and nephelometric techniques are widely used to assess turbidity and stability of nanoemulsions (Section 5.6.6).

5.6.6 Turbidimetry

Turbidimetry and nephelometry are used to assess the turbidity of nanoemulsions, which correlates with particle size and stability [31]. Turbidimetry measures the loss of light intensity as it passes through the emulsion, whereas nephelometry measures the intensity of scattered light. These techniques provide a quick and non-invasive means of evaluating particle size distribution and monitoring instability, such as particle aggregation or phase separation, over time.

5.6.7 Encapsulation Efficiency

The encapsulation efficiency of active ingredients within nanoemulsions is critical for ensuring their therapeutic or cosmetic effectiveness. Various methods are used to evaluate this parameter [2].

5.6.7.1 *Chromatographic Methods (HPLC and GC)*

High-performance liquid chromatography (HPLC) and gas chromatography (GC) are essential analytical techniques for determining the encapsulation efficiency of active compounds in nanoemulsions [32, 33]. These methods allow for the separation and quantification of active ingredients, providing precise data on the percentage that remains encapsulated versus what has leaked or degraded [34].

5.6.7.2 *Spectroscopic Methods (Ultraviolet–Visible Spectroscopy (UV-Vis) and FTIR)*

Spectroscopic techniques like UV-Vis and Fourier transform infrared (FTIR) spectroscopy are also used to assess encapsulation efficiency. UV-Vis spectroscopy is commonly used to analyze the optical properties and concentration of nanoemulsions. By measuring absorbance or light transmission at specific wavelengths, this technique helps determine the concentration of both the nanoemulsion and its encapsulated active ingredients [13]. Additionally, UV-Vis spectroscopy may provide insights into structural properties such as particle size and distribution by tracking shifts in absorption peaks. UV-Vis monitors changes in absorbance related to the encapsulated compounds, while FTIR analyzes molecular interactions between active ingredients and emulsifying agents by examining their vibrational spectra [35].

5.6.7.3 *pH and Conductivity Measurements*

The stability and functionality of a nanoemulsion can also be influenced by its pH and ionic strength. A pH meter is used to monitor the acidity or alkalinity of the nanoemulsion, which can impact its stability and shelf life [36]. Additionally, conductivity measurements are useful for assessing the ionic strength, providing insights into the dispersion stability of the nanoemulsion, as an increase in ionic strength can indicate potential aggregation or destabilization of the system.

5.6.8 *In Vitro* Permeation and Bioavailability Studies

5.6.8.1 *In Vitro Release Studies*

In vitro release studies help measure the rate and extent of active ingredient release from the nanoemulsion. This is essential for understanding the controlled release potential of the nanoemulsion in various applications, such as cosmetics, food, or drug delivery. These studies simulate conditions under which the nanoemulsion will release its active components, providing important data on how the nanoemulsion behaves under physiological conditions.

5.6.8.2 *Cell Culture Studies*

Cell culture studies evaluate the bioavailability and efficacy of nanoemulsions in biological systems. These studies assess how effectively nanoemulsions deliver active ingredients to target cells, as well as their cytotoxicity and therapeutic or cosmetic potential. By using biological models, researchers can optimize nanoemulsion formulations for specific applications, such as drug delivery or skincare [37].

5.7 Nanoemulsion for Stable Green Cosmetics

The shift from synthetic to natural-based cosmetics is driven by increased consumer awareness of harmful chemicals, growing health concerns related to skin sensitivity, and a preference for sustainable, eco-friendly products [6, 38]. The unique properties of nanoemulsions make them highly suitable for various applications in industries like cosmetics and pharmaceuticals. Nanoemulsions can be easily absorbed and penetrate the skin depending on factors such as droplet size and the degree of ionization between the emulsion and the skin. Moreover, rising consumer demand for more advanced, effective cosmetic products has led industries to invest significantly in research and development [11]. Cosmetic formulations are increasingly focusing on green ingredients aimed at skin protection, including sunscreens, anti-ageing treatments, anti-wrinkle creams, brightening products, and anti-acne solutions. They are often derived from plant or microbial sources. Nanoemulsions contribute to sustainability by improving product efficacy, reducing waste, minimizing the use of toxic ingredients, and supporting the development of eco-friendly products across various industries [38].

5.8 Green Nanoemulsions

The development of environmentally friendly, sustainable, and effective nano-scale emulsions can be achieved by integrating green chemistry principles with nanotechnology, resulting in the creation of green nanoemulsions. Green chemistry focuses on designing chemical processes and products that minimize the use of hazardous substances, emphasizing renewable feedstocks, improved energy efficiency, and safer chemicals. For nanoemulsions, green chemistry involves using renewable raw materials, biodegradable surfactants, and non-toxic solvents, thereby enhancing both functionality and sustainability while reducing environmental impact.

Kumar *et al.*, [15]; Nikolaev *et al.*, [39] have demonstrated demonstrated the use of green chemistry concepts in creating biodegradable surfactants for nanoemulsions. The researchers produced surfactants from naturally occurring fatty acids and sugars to create stable nanoemulsions without compromising safety or environmental impact [1]. Green chemistry principles such as prevention, atom economy, and design for degradation are central to developing green nanoemulsions. By avoiding hazardous substances and maximizing the use of renewable green materials, green nanoemulsions contribute to reducing pollution and enhancing product safety [9, 40, 41].

The biodegradability of green nanoemulsions is a critical factor, as it helps limit environmental persistence and bioaccumulation by breaking down into harmless byproducts [42]. Mitigating the long-term environmental impact requires biodegradability, as stated by Tadros [19]. Additionally, green nanoemulsions emphasize the use of renewable resources to support sustainable practices and resource conservation, such as plant-derived polymers. Salvia-Trujillo *et al.*, [1] state that the long-term sustainability and eco-friendliness of formulations are ensured by their dependence on renewable raw resources. Energy-efficient formulation methods for green nanoemulsions, such as ultrasonication and high-pressure homogenization emphasized by Sharma *et al.* [64], further reduce their overall environmental impact.

5.9 Principles of Green Nanoemulsions

The following are the principles of green chemistry; green nanoemulsions are developed and used in a sustainable and environmentally friendly manner.

5.9.1 Use of Renewable Resources

Green nanoemulsions incorporate renewable feedstocks, such as plant-based oils (e.g., soybean oil and sunflower oil) and bio-based surfactants (e.g., lecithin, saponins, and microbial surfactants like surfactin and rhamnolipid, as well as alkyl polyglycosides). By reducing reliance on non-renewable, petroleum-based materials, these formulations enhance the biodegradability of the final product, contributing to environmental sustainability [7].

5.9.2 Safer Solvents

The selection of solvents plays a critical role in green nanoemulsion formulations. Water, ethanol, and supercritical CO_2 are preferred for their low toxicity, non-volatility, and minimal environmental impact. These solvents replace traditional organic solvents, aligning with the green chemistry principle of using safer solvents and auxiliaries, thereby minimizing potential harm to both users and the environment.

5.9.3 Energy Efficiency

Green nanoemulsions emphasize energy-efficient production methods. Techniques such as spontaneous emulsification, phase inversion temperature, and emulsion inversion point are favored because of their lower energy consumption compared to high-energy methods like ultrasonication and high-pressure homogenization [23]. These low-energy emulsification techniques reduce the environmental footprint of the manufacturing process.

5.9.4 Design for Degradation

To promote environmental sustainability, green nanoemulsions are designed for easy degradation [21]. This involves selecting components that break down into non-toxic by- products, reducing environmental persistence and pollution. This design principle ensures that nanoemulsions have minimal long-term impact on ecosystems, aligning with the goals of green chemistry.

5.10 Formulation of Green Nanoemulsions

Green nanoemulsions are formulated using sustainable, eco-friendly materials and processes. The primary ingredients—oils, surfactants, and solvents—are selected for their low toxicity, renewability, and biodegradability. Additionally, the preparation techniques are designed to be both environmentally responsible and energy-efficient [4, 5].

5.10.1 Oils, Both Plant-Based and Natural

Soybean oil is widely utilized because it is renewable and biocompatible [1, 44]. According to McClements (2012), coconut oil is made from renewable resources and provides exceptional stability. Olive oil is well-known for being biodegradable and having nutritional advantages [57]. Sunflower oil is environmentally friendly and has strong emulsification qualities [13]. To encapsulate hydrophobic active components and ensure biocompatibility and biodegradability, these oils function as the lipid phase in nanoemulsions.

5.10.2 Bio-Based or Natural Surfactants

Lecithin, a naturally occurring surfactant extracted from soybeans, is highly biocompatible and efficient in emulsification [46]. Saponins, derived from plants like soapnuts, are biodegradable surfactants that are both safe for the environment and effective for emulsification. Fatty acid esters, made from organic fatty acids, are another eco-friendly surfactant option. Biocompatible proteins such as casein and whey protein naturally stabilize nanoemulsions. Surfactants play a crucial role in lowering interfacial tension between the oil and water phases, promoting the production of stable nanoemulsions [47, 48].

5.10.3 Eco-Friendly/Green Solvents

Water is the most environmentally benign solvent and is commonly used as the continuous phase in nanoemulsions. Ethanol, a biodegradable solvent with low toxicity, is often paired with water to enhance solubility [49]. Supercritical CO_2 solvent is perfect for green nanoemulsions because of its special qualities, which include low toxicity, non-flammability, and ease of recovery [50, 51]. These eco-friendly solvents replace traditional organic solvents, adhering to the principles of green chemistry by reducing environmental impact and toxicity [63].

5.11 Applications

Verma *et al.*, 2019 developed a green nanoemulsion system using essential oils as both the active ingredient and the oil phase [51]. This approach not only ensures the safety and environmental sustainability of the formulation but also maximizes the medicinal potency of the essential oils [45]. The researchers highlighted the antibacterial and antioxidant properties of these nanoemulsions, making them highly suitable for applications in cosmetic formulations and food preservation [52]. As regulations tighten and public awareness of harmful chemicals grows, the industry is increasingly moving toward green nanoemulsions. The enhanced performance characteristics of green nanoemulsions—such as increased stability and bioavailability—further justify their adoption [52, 53]. The study by Zhang *et al.* [69] reported that green nanoemulsions formulated with biodegradable polymers and natural oils exhibited superior stability and controlled release properties compared to conventional nanoemulsions [54]. The study demonstrated the potential of these green nanoemulsions in drug delivery applications, providing prolonged therapeutic effects while Minimizing side effects [55], Sánchez-López *et al.*, looked at the encapsulation of beneficial substances like resveratrol and curcumin using green nanoemulsions in the context of applications in the food sector [56]. The antioxidant activity and improved bioavailability demonstrated by the data underscore the potential of green nanoemulsions in functional food compositions [57]. In the food sector, natural oils such as olive oil and essential oils like lavender are employed to improve food texture, extend shelf life, and encapsulate bioactive compounds. These oils are ideal for food applications due to their non-toxicity and additional health benefits [58, 59]. Active compounds like vitamins and antioxidants are delivered by the cosmetic industry using biodegradable surfactants that come from natural sources like lecithin and glyceryl monostearate. These ingredients enhance skin hydration and cosmetic product stability. Green nanoemulsions are a safe and efficient method of medication delivery because they use biodegradable and biocompatible ingredients to increase the solubility and bioavailability of poorly water-soluble medicines [59]. One prominent example of how green A notable example of green nanoemulsions enhancing oral bioavailability is the development of self-microemulsifying drug delivery systems, which follow green chemistry principles [59, 60]. Silva *et al.* [70] explored the stabilization of green nanoemulsions for usage in food applications using chitosan, a biodegradable polymer. The study demonstrated that chitosan-based nanoemulsions improve the stability and distribution of nutraceuticals [1]. Similarly, Baloch *et al.*, [60]; Kouassi *et al.*, [61] used aloe vera and tea tree oil as natural surfactants and active ingredients, showcasing the application of green nanoemulsions in the cosmetics industry.

5.12 Case Studies on Green Nanoemulsions

5.12.1 Single Surfactant-Based Nanoemulsion System

This case study demonstrates the potential of surfactin, a bio-emulsifier, in skincare cosmetics, where a nanoemulsion was formulated exclusively using this biosurfactant as the emulsifying agent. Natural ingredients such as coconut oil as the base oil, plant-derived essential oil as the bioactive component, and surfactin as the emulsifier were employed to create a stable cosmetic nanoemulsion through a high-energy ultrasonication method. The stability of the nanoemulsion was optimized with an oil-to-surfactant ratio of 7.4:1 (% w/w), an ultrasonication amplitude of 40%, and a 4.5-minute ultrasonication period. The resulting nanoemulsion had an average droplet size of 170 ± 12 nm, a PDI of 0.17 ± 0.01, and a zeta potential of 56 ± 0.5 mV. These nanoemulsions not only exhibited excellent stability but also demonstrated significant antibacterial and antioxidant activity against *Staphylococcus aureus* [1].

A highly stable nanoemulsion was successfully formulated using surfactin as a green emulsifier, maintaining its stability for an impressive duration of 750 days. This nanoemulsion effectively encapsulates excipients such as vitamin C, vitamin A, and TiO_2. It demonstrated adaptability across various oil phases, including coconut oil, olive oil, and castor oil. The excipient-loaded nanoemulsions exhibited potent antibacterial activity against *E. coli*, *S. aureus*, and *B. subtilis*, highlighting their potential for targeted antibacterial applications. These findings offer promising prospects for encapsulating and delivering bioactive compounds with customized functionality for diverse industrial applications [32].

5.12.2 Two/Three Biosurfactant Systems: Improved Stability Mechanisms

The potential mechanisms for improved stability in two- or three-biosurfactant nanoemulsion systems were explored by studying the compatibility of different biosurfactants, including plant-derived saponin and microbial-derived rhamnolipid and surfactin, in generating stable O/W nanoemulsions using high-energy ultrasonication. Some combinations of mixed biosurfactants exhibited stability comparable to that of single-biosurfactant systems. The highest stability was observed for mixtures containing rhamnolipid and surfactin in a 75:25 ratio, resulting in an HDD of 168.9 nm and a zeta potential of −77.43 ± 2.11 mV [62]. However, other combinations, such as rhamnolipid with saponin and surfactin with saponin, demonstrated lower stability, with the surfactin-saponin mixture only achieving stability at a 75:25 ratio. Both combinations that included microbial biosurfactants resulted in stable nanoemulsions, where the unique structure of saponin played a crucial role in enhancing stability. Furthermore, antimicrobial and scavenging tests indicated that mixed-surfactant systems exhibited comparable effectiveness to single-surfactant emulsions. This suggests that the optimal synergy for stabilization is a complex function of the types and ratios of biosurfactants used [62].

5.13 Conclusion and Future Perspectives

Green nanoemulsions are a breakthrough in sustainable technology, offering eco-friendly products for the food, cosmetic, and pharmaceutical industries. These nanoemulsions deliver enhanced performance, safety, and environmental benefits by applying the principles of green chemistry. This shift toward green nanoemulsions is a proactive response to consumer and regulatory demands and represents a step toward a more sustainable future. Continued research and development in this field will enable the creation of innovative and environmentally friendly products, promoting sustainable technologies and green engineering. The future of green nanoemulsions depends on the continual research and development of environmentally friendly ingredients, scalable manufacturing techniques, and broader industrial applications. Advances in green chemistry and nanotechnology will drive the development of next-generation green nanoemulsions, offering environmentally friendly options for medications, food, and cosmetics.

AI disclosure statement: The author acknowledges the use of AI-based tools for minor language editing or grammatical correction. However, all intellectual contributions, analyses, and conclusions presented in this chapter are entirely the author's own.

Bibliography

1. Salvia-Trujillo, L., Qian, C., Martín-Belloso, O., McClements, D.J., Influence of particle size on lipid digestion and β-carotene bioaccessibility in emulsions and nanoemulsions. *Food Chem.*, 141, 2, 1472–1480, 2013. https://doi.org/10.1016/j.foodchem.2013.03.050.

2. Jaiswal, M., Dudhe, R., Sharma, P.K., Nanoemulsion: an advanced mode of drug delivery system. *3 Biotech,* 5, 2, 129–143, Apr. 01, 2015, https://doi.org/10.1007/s13205-014-0214-0.
3. Kadziński, M., *et al.*, Co-constructive development of a green chemistry-based model for the assessment of nanoparticles synthesis. *Eur. J. Oper. Res.*, 264, 2, 472–490, Jan. 2018, doi: 10.1016/j.ejor.2016.10.019.
4. Verma, A., Gautam, S., Bansal, K., Prabhakar, N., Rosenholm, J., Green Nanotechnology: Advancement in Phytoformulation Research. *Medicines*, 6, 1, 39, Mar. 2019, doi: 10.3390/medicines6010039.
5. Dinesh, M.K.S., *et al.*, Review On Green Nanotechnology. *J. Pharm. Negat. Results*, 13, S09, 10975–10979, 2022. doi: 10.47750/pnr.2022.13.S09.1283.
6. Singh, R.D., Kapila, S., Ganesan, N.G., Rangarajan, V., A review on green nanoemulsions for cosmetic applications with special emphasis on microbial surfactants as impending emulsifying agents. *J. Surfactants Deterg.*, 25, 3, 433–451, May 01, 2022, John Wiley and Sons Inc. https://doi.org/10.1002/jsde.12571.
7. Ashaolu, T.J., Nanoemulsions for health, food, and cosmetics: a review. Aug. 01, 2021, Springer Science and Business Media Deutschland GmbH. doi: 10.1007/s10311-021- 01216-9.
8. Graça, A., Martins, A. M., Pinto, P., Ribeiro, H. M., Marto, J., Combining protection with skin health: *In vivo* studies of an innovative gelatin/tannic acid-based hydrogel patch to prevent PPE-related skin lesions. *Int. J. Pharm.*, 650, 123731, 2024. https://doi.org/10.1016/j.ijpharm.2023.123731.
9. Tayal, S., Tiwari, P., Pratap Singh, U., Dubey, S., Vishwakarma, R., A Review on Formulation, Characterization and Applications of Nanoemulsion. *Asian J. Appl. Sci. Technol.*, 08, 01, 67–84, 2024, doi: 10.38177/ajast.2024.8106.
10. Najan, A., Review on Nano-Emulsion. 2023, doi: 10.36648/2321-547X.23.11.011.
11. Souto, E.B., Cano, A., Martins-Gomes, C., Coutinho, T.E., Zielińska, A., Silva, A.M., Microemulsions and Nanoemulsions in Skin Drug Delivery. *Bioengineering,* 9, 4, 158. Apr. 01, 2022, MDPI. https://doi.org/10.3390/bioengineering9040158.
12. Bolzinger, M.A., Briançon, S., Pelletier, J., Chevalier, Y., Penetration of drugs through skin, a complex rate-controlling membrane. *Curr. Opin. Colloid Interface Sci.,* 17, 3, 171–176, Jun. 2012, https://doi.org/10.1016/j.cocis.2012.02.001.
13. Preeti, *et al.*, Nanoemulsion: An Emerging Novel Technology for Improving the Bioavailability of Drugs. *J. Nanomater.*, 2023, 6640103, 2023, Hindawi Limited. https://doi.org/10.1155/2023/6640103.
14. Yousefpoor, Y., Esnaashari, S.S., Baharifar, H., Mehrabi, M., Amani, A., Current challenges ahead in preparation, characterization, and pharmaceutical applications of nanoemulsions. *Wiley Online Library*, Nov. 01, 2023, John Wiley and Sons Inc. https://doi.org/10.1002/wnan.1920.

15. Kumar, M., Bishnoi, R.S., Shukla, A.K., Jain, C.P., Techniques for formulation of nanoemulsion drug delivery system: A review. *Prev. Nutr. Food Sci.*, 24, 3, 225–231, 2019. https://doi.org/10.3746/pnf.2019.24.3.225.
16. Smail, S.S., Ghareeb, M.M., Omer, H.K., Al-Kinani, A.A., Alany, R.G., Studies on surfactants, cosurfactants, and oils for prospective use in formulation of ketorolac tromethamine ophthalmic nanoemulsions. *Pharmaceutics*, 13, 4, 467, Apr. 2021, https://doi.org/10.3390/pharmaceutics13040467.
17. Sharma, V., Saxena, P., Singh, D.L., Singh, P., Self Emulsifying Drug Delivery System: A novel approach. *J. Pharm. Biomed. Sci.*, 1, 1, 1–13, 2011, [Online]. Available: https://www.researchgate.net/publication/258026600.
18. Azeem, A., Rizwan, M., Ahmad, F.J., Khar, R.K., Talegaonkar, S., Components Screening and Influence of Surfactant and Cosurfactant on Nanoemulsion Formation. 2009.
19. Tadros, A.B., Hannah Y.W., Monica M., Breast Cancers of Special Histologic Subtypes Are Biologically Diverse. *Ann. Surg. Oncol.*, 25, 11, 3158–64, 2018. https://doi.org/10.1245/s10434-018-6687-z.
20. Mushtaq, A., *et al.*, Recent insights into Nanoemulsions: Their preparation, properties and applications. *Food Chem.*, X, 18, 100684, Jun. 30, 2023, Elsevier Ltd. doi: 10.1016/j.fochx.2023.100684.
21. Singh, I.R. and Pulikkal, A.K., Preparation, stability and biological activity of essential oil-based nano emulsions: A comprehensive review. *OpenNano*, 2, 100066, Nov. 01, 2022, Elsevier Inc. doi: 10.1016/j.onano.2022.100066.
22. Al-Sakkaf, M.K. and Onaizi, S.A., Effects of emulsification factors on the characteristics of crude oil emulsions stabilized by chemical and Biosurfactants: A review. *Fuel*, 361, 130604, Apr. 01, 2024, Elsevier Ltd. doi: 10.1016/j.fuel.2023.130604.
23. Komaiko, J.S. and Mcclements, D.J., Formation of Food-Grade Nanoemulsions Using Low-Energy Preparation Methods: A Review of Available Methods. *Compr. Rev. Food Sci. Food Saf.*, 15, 2, 331–352, Mar. 2016, doi: 10.1111/1541-4337.12189.
24. Ragheb, R. and Nobbmann, U., Multiple scattering effects on intercept, size, polydispersity index, and intensity for parallel (VV) and perpendicular (VH) polarization detection in photon correlation spectroscopy. *Sci. Rep.*, 10, 1, 1–11, Dec. 2020, doi: 10.1038/s41598-020-78872-4.
25. Faria, M., *et al.*, Minimum information reporting in bio–nano experimental literature. *Nat. Nanotechnol.*, 13, 9, 777–785, Sep. 2018, doi: 10.1038/s41565-018-0246-4.
26. Klang, V., Matsko, N.B., Valenta, C., Hofer, F., Electron microscopy of nanoemulsions: An essential tool for characterisation and stability assessment. Feb. 2012, doi: 10.1016/j.micron.2011.07.014.
27. Ganta, S. and Amiji, M., Coadministration of paclitaxel and curcumin in nanoemulsion formulations to overcome multidrug resistance in tumor cells. *Mol. Pharm.*, 6, 3, 928–939, Jun. 2009, doi: 10.1021/mp800240j.

28. Seki, J., Sonoke, S., Saheki, A., Fukui, H., Sasaki, H., Mayumi, T., A nanometer lipid emulsion, lipid nano-sphere (LNS®), as a parenteral drug carrier for passive drug targeting. *Int. J. Pharm.*, 273, 1–2, 75–83, Apr. 2004, doi: 10.1016/j.ijpharm.2003.12.022.
29. Espitia, P.J.P., Fuenmayor, C.A., Otoni, C.G., Nanoemulsions: Synthesis, Characterization, and Application in Bio-Based Active Food Packaging. *Compr. Rev. Food Sci. Food Saf.*, 18, 1, 264–285, Jan. 2019, doi: 10.1111/1541-4337.12405.
30. Souza, I.D.L., Saez, V., Mansur, C.R.E., Lipid nanoparticles containing coenzyme Q10 for topical applications: An overview of their characterization. Oct. 01, 2023, Elsevier B.V. doi: 10.1016/j.colsurfb.2023.113491.
31. Bardsley, R., Gardner, G., Tse, H.M., Fraker, C.A., Optical sensor arrays designed for guided manufacture of perfluorocarbon nanoemulsions with a non-synthetic stabilizer. *Acta Biomater.*, 136, 558–569, Dec. 2021, doi: 10.1016/j.actbio.2021.09.038.
32. Ganesan, N.G., Singh, R.D., Rangarajan, V., Investigating the influence of excipients on the stability of biocosmetic nanoemulsions formulated with surfactin as an emulsifier. *Biomass Convers. Biorefin.*, 2023, doi: 10.1007/s13399-023-05080-1.
33. Rajan, V.H. and Rajan, M.V., Review Article Development and validation of HPLC method-A Review. 2015, [Online]. Available: www.ijcrp.com.
34. Rangarajan, V. and Clarke, K.G., Towards bacterial lipopeptide products for specific applications — a review of appropriate downstream processing schemes. Dec. 01, 2016, Elsevier Ltd. doi: 10.1016/j.procbio.2016.08.026.
35. Crupi, V., Ficarra, R., Guardo, M., Majolino, D., Stancanelli, R., Venuti, V., UV-vis and FTIR-ATR spectroscopic techniques to study the inclusion complexes of genistein with β-cyclodextrins. *J. Pharm. Biomed. Anal.*, 37, 1, 110–117, May 2007, doi: 10.1016/j.jpba.2007.01.054.
36. Khalid, A., Arshad, M.U., Imran, A., Haroon Khalid, S., Shah, M.A., Development, stabilization, and characterization of nanoemulsion of vitamin D3-enriched canola oil. *Front. Nutr.*, 10, 1205200, 2023, doi: 10.3389/fnut.2023.1205200.
37. Krasteva, G., Georgiev, V., Pavlov, A., Recent applications of plant cell culture technology in cosmetics and foods. Mar. 01, 2021, Wiley-VCH Verlag. doi: 10.1002/elsc.202000078.
38. Marzuki, N.H.C., Wahab, R.A., Hamid, M.A., An overview of nanoemulsion: Concepts of development and cosmeceutical applications. *J. Biomater. Sci.*, Polymer Edition, 30, 10, 875–894, 2019, Taylor and Francis Ltd. doi: 10.1080/13102818.2019.1620124.
39. Nikolaev, B., Yakovleva, L., Fedorov, V., Li, H., Gao, H., Shevtsov, M., Nano- and Microemulsions in Biomedicine: From Theory to Practice. Jul. 01, 2023, Multidisciplinary Digital Publishing Institute (MDPI). doi: 10.3390/pharmaceutics15071989.

40. Schreiner, T.B., *et al.*, Formulation and optimization of nanoemulsions using the natural surfactant saponin from Quillaja bark. *Molecules*, 25, 7, 1538, 2020, doi: 10.3390/molecules25071538.
41. Jiang, T., Liao, W., Charcosset, C., Recent advances in encapsulation of curcumin in nanoemulsions: A review of encapsulation technologies, bioaccessibility and applications. *J. Dispersion Sci. Technol.*, 43, 7, 1061–1078, Jun. 01, 2020, Elsevier Ltd. doi: 10.1016/j.foodres.2020.109035.
42. Tiwari, U., Ganesan, N.G., Junnarkar, J., Rangarajan, V., Toward the formulation of bio-cosmetic nanoemulsions: from plant-derived to microbial-derived ingredients. *J. Dispersion Sci. Technol.*, 36, 7, 1061–1078, 2022, doi: 10.1080/01932691.2020.1847664.
43. Sharan, P. and Singh, S., Applications of green solvents in extraction of phytochemicals from medicinal plants: A review. 2018, [Online]. Available: https://www.researchgate.net/publication/340635250.
44. Kawaguchi, M., Silicone oil emulsions stabilized by polymers and solid particles. *J. Colloid Interface Sci.*, 473, 1–13, Jul. 01, 2016, Elsevier. doi: 10.1016/j.cis.2015.06.005.
45. McClements, D.J., Nanoemulsions versus microemulsions: Terminology, differences, and similarities. *Soft Matter*, 8, 6, 1719–1729, 2012. https://doi.org/10.1039/C2SM06903B.
46. Ismail, N.L., Shahruddin, S., Othman, J., Perspective Chapter: Overview of Bio-Based Surfactant - Recent Development, Industrial Challenge, and Future Outlook, in: *A. K. Dutta (Ed.), Surfactants and Detergents - Updates and New Insights*, pp. 25–45, IntechOpen, 2022. https://doi.org/10.5772/intechopen.100542.
47. Smułek, W., *et al.*, Whey proteins as a potential co-surfactant with aesculus hippocastanum l. As a stabilizer in nanoemulsions derived from hempseed oil. *Molecules*, 26, 19, 5856, Oct. 2021, doi: 10.3390/molecules26195856.
48. Liu, Q., Huang, H., Chen, H., Lin, J., Wang, Q., Food-grade nanoemulsions: Preparation, stability and application in encapsulation of bioactive compounds. *Molecules*, 24, 23, 4242, Nov. 19, 2019, MDPI AG. doi: 10.3390/molecules24234242.
49. Almohasin, J.A., Balag, J., Miral, V.G., Moreno, R.V., Tongco, L.J., Lopez, E.C.R., Green Solvents for Liquid–Liquid Extraction: Recent Advances and Future Trends †. *Eng. Proc.*, 45, 1, 2023, doi: 10.3390/ASEC2023-16278.
50. Welton, T., Solvents and sustainable chemistry. Nov. 08, 2015, Royal Society of London. doi: 10.1098/rspa.2015.0502.
51. Verma, A., Gautam, S., Bansal, K., Prabhakar, N., Rosenholm, J., Green Nanotechnology: Advancement in Phytoformulation Research. *Medicines*, 6, 1, 39, MDPI, 2019.

52. Azmi, N.A.N., Elgharbawy, A.A.M., Motlagh, S.R., Samsudin, N., Salleh, H.M., Nanoemulsions: Factory for food, pharmaceutical and cosmetics. Sep. 01, 2019, MDPI AG. doi: 10.3390/pr7090617.
53. Gupta, R., *et al.*, Recent progress on nanoemulsions mediated pesticides delivery: Insights for agricultural sustainability. *Plant Nano Biol.,* 8, 100073. May 01, 2024, Elsevier B.V. doi: 10.1016/j.plana.2024.100073.
54. Fenibo, E.O., Ijoma, G.N., Nurmahomed, W., Matambo, T., The Potential and Green Chemistry Attributes of Biopesticides for Sustainable Agriculture. *Sustainability*, 14, 21, 14417, Nov. 2022, doi: 10.3390/su142114417.
55. Wilson, R.J., Li, Y., Yang, G., Zhao, C.X., Nanoemulsions for drug delivery. *Particuology*, 45, 85–97, May 2022, doi: 10.1016/j.partic.2021.05.009.
56. Sánchez-López, E., *et al.*, Current applications of nanoemulsions in cancer therapeutics. Nanomaterials, 9, 6, 821, Jun. 01, 2019, MDPI AG. doi: 10.3390/nano9060821.
57. Vieira, I.R.S., Tessaro, L., Lima, A.K.O., Velloso, I.P.S., Conte-Junior, C.A., Recent Progress in Nanotechnology Improving the Therapeutic Potential of Polyphenols for Cancer. Jul. 01, 2023, Multidisciplinary Digital Publishing Institute (MDPI). doi: 10.3390/nu15143136.
58. Wang, S., Cheng, Y., Wang, J., Ding, M., Fan, Z., Antioxidant Activity, Formulation, Optimization and Characterization of an Oil-in-Water Nanoemulsion Loaded with Lingonberry (Vaccinium vitis-idaea L.) Leaves Polyphenol Extract. *Foods*, 12, 23, 4256, Dec. 2023, doi: 10.3390/foods12234256.
59. Taghvaei, M. and Jafari, S.M., Application and stability of natural antioxidants in edible oils in order to substitute synthetic additives. *J. Food Sci. Technol.,* 52, 1288–1296, Mar. 01, 2015, Springer. doi: 10.1007/s13197-013-1080-1.
60. Baloch, J., *et al.*, Self-nanoemulsifying drug delivery system (Snedds) for improved oral bioavailability of chlorpromazine: *In vitro* and *in vivo* evaluation. *Medicina (Lithuania)*, 55, 5, 210, May 2019, doi: 10.3390/medicina55050210.
61. Kouassi, M.C., Grisel, M., Gore, E., Multifunctional active ingredient-based delivery systems for skincare formulations: A review. *Colloids and Surfaces B: Biointerfaces,* 217, 112676, Sep. 01, 2022, Elsevier B.V. doi: 10.1016/j.colsurfb.2022.112676.
62. Ganesan, N.G., Singh, R.D., Dwivedi, D., Rangarajan, V., Synergy evaluation between diverse biosurfactants toward the formulation of green oil-in-water nanoemulsions by ultrasonication method. *J. Clean. Prod.*, 400, 136735, May 2023, doi: 10.1016/j.jclepro.2023.136735.
63. Tiwari, U., Ganesan, N.G., Junnarkar, J., Rangarajan, V., Toward the formulation of bio-cosmetic nanoemulsions: from plant-derived to microbial-derived ingredients. *J. Dispersion Sci. Technol.*, 43, 7, 1061–1078, 2020. https://doi.org/10.1080/01932691.2020.1847664.
64. Sharma, S., Sridhar, S., McIntosh, A., *et al.*, Periodontal Therapy and Treatment of Hypertension-Alternative to the Pharmacological Approach. A Systematic Review and Meta-Analysis. *Pharmacol. Res.*, 166, 105511, April 2021. https://doi.org/10.1016/j.phrs.2021.105511.

65. Silverman, J.R., Exploring sustainability metrics in general chemistry using intensive and extensive properties of matter. *J. Chem. Educ.*, 98, 9, 2741–2745, 2021. https://doi.org/10.1021/acs.jchemed.1c00113.
66. Hashem, A.S., Ramadan, M M., Abdel-Hady, A. A. A., Sut, S., Maggi, F., Dall'Acqua, S., Pimpinella anisum essential oil nanoemulsion toxicity against Tribolium castaneum? Shedding light on its interactions with aspartate aminotransferase and alanine aminotransferase by molecular docking. *Molecules*, 25, 20, 4841, 2020. https://doi.org/10.3390/molecules25204841.
67. Guzmán, E., Fernández-Peña, L., Rossi, L., Bouvier, M., Ortega, F., Rubio, R.G., Nanoemulsions for the encapsulation of hydrophobic actives. *Cosmetics*, 8, 2, 45, 2021. https://doi.org/10.3390/cosmetics8020045.
68. Lewińska, A., Domżał-Kędzia, M., Maciejczyk, E., Łukaszewicz, M., Bazylińska, U., Design and engineering of 'green' nanoemulsions for enhanced topical delivery of Bakuchiol achieved in a sustainable manner: A novel eco-friendly approach to bioretinol. *Int. J. Mol. Sci.*, 22, 18, 10091, 2021. https://doi.org/10.3390/ijms221810091.
69. Zhang, Y., Chen, J., Zhang, R., McClements, D.J., Development of biopolymer-based green nanoemulsions for enhanced stability and controlled release of bioactives. *Food Hydrocoll.*, 126, 107437, 2022.
70. Silva, H.D., Cerqueira, M.A., Vicente, A.A., Stabilization of green nanoemulsions for food and nutraceutical applications: Physicochemical aspects and future perspectives. *Food & Function*, 9, 3, 1212–1226, 2018.

6

Cutting-Edge Eco-Conscious Technologies for Glycan Separation

M. Aniskumar, C. Santhaseelan, K. Keerthiga, G. Jeyashree and M. A. Sundaramahalingam*

Department of Biotechnology, V. S. B. Engineering College, Karur, Tamil Nadu, India

Abstract

This chapter explores the revolutionary developments in environmentally and financially advantageous sustainable glycan separation techniques. Many green technologies are included in the debate, including membrane filtration, adsorption, chromatography, biocatalysis, green solvents, and electrochemical methods. These techniques minimize environmental effect while providing notable gains in selectivity, sensitivity, and throughput. The efficiency and sustainability of glycan separation procedures are further improved by the integration of nanotechnology, artificial intelligence, and sophisticated manufacturing techniques. The chapter emphasizes the possibility for industrial adoption through creative ideas and teamwork, despite obstacles including scalability, economic feasibility, and regulatory compliance. Future developments suggest that green technology will continue to advance, with a focus on nanomaterials, microfluidics, and biomimetic techniques. To encourage the adoption of these environmentally beneficial technologies, the chapter ends with a plea for greater education and awareness. This will ultimately help to improve glycomics and promote global sustainability. When it comes to glycan analysis and production, this thorough review is an invaluable tool for researchers and industry experts looking to adopt more environmentally friendly processes.

Keywords: Glycan, green technologies, sustainability, separation techniques, carbohydrates

**Corresponding author*: Sundar7mps@gmail.com

Senthilkumar Rathnasamy, Vivek Rangarajan and Umile Gianfranco Spizzirri (eds.) Green Engineering and Sustainable Technology: Tools and Techniques, (161–194) © 2026 Scrivener Publishing LLC

6.1 Introduction

Glycans are a broad class of molecules that include glycolipids, glycoproteins, and glycosaminoglycans, all of which have unique biological roles and topologies. These molecules are critical for biochemical signaling pathways, immune response modulation, and cellular recognition. For example, glycoproteins on the surface of cells act as receptors for signaling chemicals and pathogens, and they also help to enable connections between cells. Furthermore, the stability, location, and functionality of proteins and lipids within cells are frequently dictated by the glycans that bind to them [1]. Beyond fundamental cellular processes, glycans are important in the contexts of health and illness. Changes in glycan structures have been connected to several diseases, including metabolic disorders and cancer, as biomarkers. Knowing these glycosyl changes helps with both disease diagnosis and the development of treatment approaches that specifically target glycan-mediated pathways [2].

Glycans are complicated molecules that require advanced analytical methods to examine. The comprehensive structural characterization and quantification of glycans in biological samples has been made possible by techniques like mass spectrometry, chromatography, and glycan microarrays, which have revolutionized the discipline. These techniques are essential for developing glycan-based therapeutics and clarifying the functions of glycans in health and illness. Glycan extraction and purification have historically involved energy-intensive procedures and chemical solvents, which worsen the environment and increase carbon footprints. These techniques not only use much energy, but they also produce waste streams that, if improperly managed, could be dangerous. The need for environmentally conscious technologies that minimize resource usage, limit waste output, and use greener solvents and processes is growing as awareness of environmental sustainability rises [3]. Traditional glycan separation methods have an adverse effect on the environment because they release greenhouse gases, deplete natural resources, and contaminate ecosystems. Additionally, there may be concerns for the environment and public health from chemical solvents that are frequently employed in glycan extraction techniques. Innovative strategies that put sustainability first without sacrificing the accuracy and effectiveness of glycan isolation and analysis are needed to meet these difficulties [4].

Green chemistry and engineering developments in recent years have opened the door for environmentally friendly glycan separation technologies. These approaches include the use of biodegradable solvents that allow effective glycan extraction with minimal environmental impact, like deep eutectic solvents (DESs) and ionic liquids (ILs). Furthermore, as sustainable alternatives to conventional procedures, membrane-based separation techniques and enzymatic methods have shown promise in lowering energy usage and waste creation. The critical need for sustainable separation technologies is highlighted by the fact that conventional glycan separation methods frequently rely on solvents and processes that present concerns to the environment and human health [5]. The goals and scope of developing environmentally sensitive technologies for glycan separation are examined in this chapter, with a focus on efficiency and sustainability in the glycan extraction and purification procedures.

The main goal of this chapter is to examine how environmentally sensitive glycan separation technologies are developing. Our goal is to examine existing techniques, draw attention to how they affect the environment, and investigate novel ideas that support sustainability without compromising the purity and usefulness of glycans. Overview of glycan structures and functions, problems with conventional glycan separation methods, emerging eco-conscious technologies, case studies and applications, and future directions and implications are the main topics of this chapter. This chapter seeks to promote innovation and stewardship in the field of biomolecular separation by addressing these goals and furthering ecologically sustainable techniques in glycan research and application.

6.2 Fundamentals of Glycan Separation

6.2.1 Types of Glycans

Complex carbohydrates called glycans, or saccharides, are essential to many biological functions. They are made up of monosaccharides joined by glycosidic linkages to form a variety of structures, from simple sugars to intricate polysaccharides. Studying the many forms and structures of glycans is crucial for learning about their uses in biotechnology and medicine. The forms of various types and structures of glycans are depicted in Figure 6.1. As the most basic type of carbohydrates, monosaccharides are made up of just one sugar molecule. They function as building blocks for more complex glycans and give rise to energy *via* metabolic pathways. Disaccharides are carbohydrates made up of a glycosidic bond connecting two monosaccharide molecules. It is frequently present in food and aids in the metabolism and storage of energy. Short chains of monosaccharide units called oligosaccharides, which usually consist of three to ten sugar molecules, are involved in immunological response, cell recognition, and prebiotic

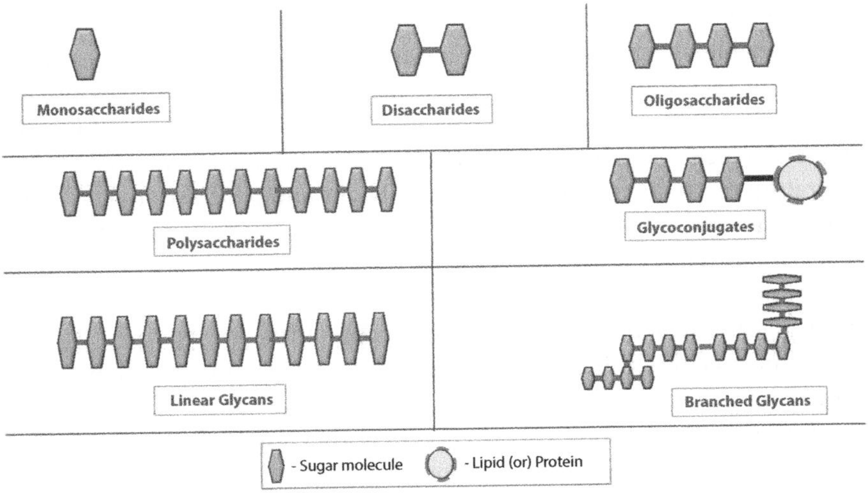

Figure 6.1 Various types of glycans.

activities in the gut. Long chains of monosaccharide units called polysaccharides, which include more than ten sugar molecules, have a variety of biological activities in addition to acting as structural elements like cellulose in plants and energy stores like glycogen in animals. Glycoconjugates are compounds made of glycans covalently bound to lipids or proteins. It plays a decisive role in cell-cell communication, signaling, and structural integrity of cells and tissues [6].

Glycans that are linear consist of a single chain of monosaccharide units connected by glycosidic linkages. Depending on their links, they can either be water-soluble or insoluble or usually form stiff structures. Branched glycans offer a lot of surface area for interactions and are frequently involved in intricate biological processes like cell signaling. Branched glycans are glycans with a branching structure where numerous chains of monosaccharides stretch from a central core. Glycoconjugates contain highly branched, complex structures called complex glycans, which, depending on how they are modified, can serve a variety of purposes. They are crucial for cellular communication, stability, and protein folding [7].

6.2.2 Glycan Structure

N-linked glycans are essential for the folding and stability of proteins. They are linked to the nitrogen atom of asparagine residues in proteins. O-linked glycans are structures that influence how proteins function and how cells communicate. They are attached to the oxygen atomof serine or threonine residues.

Long, unbranched polysaccharides made up of repeating disaccharide units and frequently sulfated are called glycosaminoglycans. Examples are chondroitin and heparan sulfates, which are necessary for extracellular matrix maintenance and cell signaling. Glycan structures are diverse and complicated, making it necessary to characterize and manipulate them using sophisticated analytical techniques. These approaches are essential to the advancement of glycoscience [8].

6.2.3 Glycan Separation Techniques

In glycoscience, glycan separation is an essential step that is required for the in-depth examination of their composition, uses, and functions. The process is made more difficult by several issues, including the limitations of the available analytical methods and the intrinsic complexity of glycan structures. For glycan research and its uses in biotechnology and medicine to advance, these issues must be resolved. Because of differences in branching patterns, linking types, monosaccharide composition, and modifications like acetylation and sulfation, glycans have a vast array of structural variants. Glycan analysis becomes tedious due to structural isomers (same molecular formula but various structures) and stereoisomers (same structural formula but different spatial arrangements). Accurately separating and identifying distinct glycan species is difficult due to this variability. Glycans linked to the same protein or lipid have slight structural differences, resulting in a population of molecules that are similar but not identical. The separation and characterization procedures are made more difficult by microheterogeneity, necessitating the use of extremely precise and sensitive analytical methods. Analytical difficulties with glycan analysis include sensitivity, specificity, sample preparation, complex matrices, and degradation. Additionally, there were technological constraints at play, including those related to instrumentation throughput, cost and accessibility, complex data analysis, and data standardization. Glycan separation issues were also influenced by a few environmental and health concerns, such as the toxicity of chemical solvents, their impact on the environment, and the generation of hazardous waste [9].

Understanding glycans' functions in biological systems and creating glycan-based technologies depend on their isolation and study. Although conventional separation techniques have been in use for many years, they have several drawbacks that make the creation of more sophisticated and environmentally friendly procedures necessary. Table 6.1 describes the inherent limitations of the widely used conventional glycan separation procedures.

Table 6.1 Traditional methods of glycan separation.

Techniques	Applications	Limitations	References
High-performance liquid chromatography	Used for separating glycans based on size, charge, and hydrophobicity	Requires extensive sample preparation and uses large volumes of organic solvents, which can be hazardous and costly	[3]
Size-exclusion chromatography	Commonly used for polysaccharides and large glycan complexes	Limited resolution for molecules of similar size and requires large sample volumes	[10]
Ion-exchange chromatography	Effective for separating acidic and basic glycans	Sensitivity to changes in pH and ionic strength can result in glycan degradation	[11]
Capillary electrophoresis	High-resolution separation of monosaccharides and oligosaccharides	Limited sample capacity and requires precise control of experimental conditions to maintain reproducibility	[12]
Polyacrylamide gel electrophoresis	Commonly used for glycoproteins and complex glycan mixtures	It is time-consuming, requires staining for visualization, and can be less effective for highly branched glycans.	[13]

(*Continued*)

Table 6.1 Traditional methods of glycan separation. (*Continued*)

Techniques	Applications	Limitations	References
Matrix-assisted laser desorption/ ionization mass spectrometer	It provides detailed structural information and is effective for high-throughput analysis.	Requires extensive sample preparation and derivatization and can suffer from ion suppression effects	[14]
Electrospray ionization mass spectrometer	Suitable for analyzing complex glycans and glycopeptides	Complex data interpretation and potential issues with ionization efficiency	[15]

These methods also include some other limitations related to environmental impacts like solvent use and energy consumption, complexity and purity in sample preparation, resolution and sensitivity of each method and reproducibility factors like variability and standardization. So, it resulted in the search for new green separation technologies, which are eco-friendly and sustainable, employing green solvents. It also ended up in developing new technological innovations, such as advanced instrumentation and computational tools for glycan separation and characterization [6].

6.3 Principles of Green Technologies in Glycan Separation

The goal of green chemistry, sometimes referred to as sustainable chemistry, is to create chemical processes and products that minimize or completely do away with the need for dangerous materials. Green chemistry approaches must be applied in glycan separation to reduce environmental effects and improve the sustainability of analytical techniques. This section highlights the advantages and possible drawbacks of applying green chemistry concepts to glycan separation methods. Various principles of green chemistry are tabulated as Table 6.2.

Table 6.2 Principles and applications of green chemistry.

Particular	Principle	Application	References
Prevention	It is better to prevent waste than to treat or clean up waste after it is created.	Developing separation techniques that minimize the generation of waste, such as using minimal solvent volumes and employing efficient sample preparation methods	[17]
Atom economy	Synthetic methods should maximize the incorporation of all materials used in the process into the final product.	Designing glycan derivatization reactions that incorporate all reactants into the final analyte, reducing by-products and improving overall efficiency	[18]
Less hazardous chemical syntheses	Wherever practicable, synthetic methods should be designed to use and generate substances that possess little or no toxicity to human health and the environment.	Using non-toxic, biodegradable solvents and reagents for glycan separation and analysis	[19]
Designing safer chemicals	Chemical products should be designed to affect their desired function while minimizing their toxicity.	Creating safer chemical derivatization agents for glycans that are effective yet non-toxic	[20]

(*Continued*)

Table 6.2 Principles and applications of green chemistry. (*Continued*)

Particular	Principle	Application	References
Safer solvents and auxiliaries	The use of auxiliary substances (e.g., solvents and separation agents) should be made unnecessary wherever possible and innocuous when used.	Replacing traditional organic solvents with greener alternatives such as water, supercritical fluids, or ionic liquids	[21]
Design for energy efficiency	Energy requirements should be recognized for their environmental and economic impacts and should be minimized. Synthetic methods should be conducted at ambient temperature and pressure.	Optimizing separation processes to operate at room temperature and pressure, reducing energy consumption	[20]
Use of renewable Feedstocks	A raw material or feedstock should be renewable rather than depleting wherever technically and economically practicable.	Sourcing reagents and materials from renewable resources for glycan separation techniques	[21]
Reduce derivatives	Unnecessary derivatization (blocking group, protection/deprotection, and temporary modification) should be avoided whenever possible.	Simplifying sample preparation by minimizing the need for chemical modifications of glycans	[22]

(*Continued*)

Table 6.2 Principles and applications of green chemistry. (*Continued*)

Particular	Principle	Application	References
Catalysis	Catalytic reagents (as selective as possible) are superior to stoichiometric reagents.	Employing catalytic processes in glycan derivatization and separation to enhance efficiency and reduce waste	[23]
Design for degradation	Chemical products should be designed so that at the end of their function they break down into innocuous degradation products and do not persist in the environment.	Ensuring that all chemicals used in glycan separation are biodegradable and do not produce harmful by-products	[20]
Real-time analysis for pollution prevention	Analytical methodologies need to be further developed to allow for real-time, in-process monitoring and control prior to the formation of hazardous substances.	Implementing real-time monitoring techniques during glycan separation to ensure processes that are within safe environmental parameters	[24]
Inherently safer chemistry for accident prevention	Substances and the form of a substance used in a chemical process should be chosen to minimize the potential for chemical accidents, including releases, explosions, and fires.	Designing separation processes that use safe chemicals and conditions to prevent accidents	[25]

Waste prevention is preferable to waste treatment or cleanup after it has been produced. Separation processes that reduce waste production can be created by, for example, using small amounts of solvent and effective sample preparation techniques. The ultimate result of synthetic methods should incorporate as many raw components as possible. It creates glycan derivatization processes that maximize efficiency by eliminating by-products and incorporating all reactants into the final analyte. Synthetic techniques should, whenever possible, be created using and producing materials with little to no toxicity to the environment and human health. Biodegradable, non-toxic reagents and solvents were used to separate and analyze glycans. Chemical products ought to be made with the least amount of toxicity possible without sacrificing the intended purpose. They are developing non-toxic but safe chemical derivatization compounds for glycans. When possible, auxiliary compounds (such as solvents and separation agents) should be avoided or used sparingly and safely. They are substituting more environmentally friendly options like water, supercritical fluids, or ILs for conventional organic solvents [16].

Energy needs should be reduced with an awareness of their effects on the economy and environment. It is best to use synthetic procedures at room temperature and pressure. It lowers energy consumption by optimizing separation processes to run at ambient temperature and pressure. Whenever it is both technically and financially feasible, a raw material or feedstock should be renewable rather than diminishing. It is obtaining ingredients and reagents for glycan separation processes from renewable resources. It is best to steer clear of needless derivatization (blocking group, protection/deprotection, and temporary alteration). They are cutting down on the requirement for glycan chemical alterations to simplify sample preparation. Stoichiometric reagents are not as good as catalytic reagents (as selective as possible). Glycan derivatization and separation using catalytic methods to increase productivity and decrease waste. Chemical products ought to be made in such a way that, upon reaching the end of their useful life, they decompose into harmless by-products and disappear from the environment. It ensures that no hazardous by-products are produced and that all chemicals employed in glycan separation are biodegradable. In order to provide real-time, in-process monitoring and control before the creation of hazardous chemicals, analytical methods must be further developed. All separation operations should be monitored in real-time to make sure they are taking place within safe environmental bounds. When using substances in a chemical process, care should be taken in selecting their shape and composition to reduce the risk of leaks, explosions, and fires.

It creates separation procedures with safe chemicals and surroundings to avoid mishaps [21].

Principles of green chemistry related to glycan separation are being applied to a variety of green solvents, including ILs, supercritical fluids, and water. It is also used in energy-efficient techniques (such as ultrasound- and microwave-assisted separation and extraction), catalytic processes (such as biocatalysis and enzymatic methods), and real-time monitoring (such as process analytical technology and in-line sensors). These applications reduce the use of hazardous chemicals and solvents, increase energy efficiency, lower operating costs, and improve safety. The main drawbacks of green technologies are their high research and development costs, which make it difficult to guarantee optimal performance in comparison to conventional methods. It could be necessary to overcome reluctance to change and provide undeniable advantages to persuade the scientific community and industry to embrace green approaches [20].

In addition to being a scientific breakthrough, the incorporation of environmentally friendly technologies into glycan separation is a big step toward sustainable practices that will help the environment in general. The numerous environmental advantages of employing green methods for glycan separation are discussed in this part, with an emphasis on waste minimization, energy conservation, and the use of safer chemicals. Significant amounts of hazardous waste, including organic solvents and reagents that are dangerous to properly dispose of, are frequently produced by traditional glycan separation techniques. Sustainable technologies work to reduce this waste by utilizing fewer reagent volumes and more environmentally friendly solvents. In order to ensure that any trash formed during the glycan separation process breaks down naturally and does not pollute the environment over time, biodegradable reagents are used in the procedure. Energy consumption of eco-friendly technology is typically lower than that of conventional techniques. Methods like extraction assisted by microwave and separation helped by ultrasound reduce energy consumption by increasing productivity and speeding up processes. In order to minimize the energy needed for heating and pressurization in traditional procedures, several green separation techniques are made to function at room temperature and pressure. Laboratory worker safety is improved and environmental contamination risk is decreased when conventional organic solvents are swapped out for non-toxic, ecologically friendly alternatives like water, supercritical CO2, or ILs. Eco-friendly technologies lessen the possibility of hazardous chemical exposure for people and wildlife by using safer chemicals. In biological and pharmaceutical applications where reagent safety and purity are crucial, this is especially relevant to glycan analysis. Glycan separation using renewable feedstocks encourages sustainability and the preservation of non-renewable resources. Using products that are plant-based or bio derived, for example, lessens the need for fossil fuels. Eco-friendly technologies frequently have procedures that optimize material utilization across the board, increasing atom economy and lowering the quantity of raw materials required for glycan separation. Eco-friendly glycan separation technologies help to minimize the total carbon footprint by utilizing greener chemicals and consuming less energy. It is imperative to reduce greenhouse gas emissions in order to mitigate climate change. Energy-efficient building designs, water conservation, garbage recycling, and other larger sustainability projects are all aided by the use of green technology in laboratories. Eco-friendly solutions make working with high-energy equipment and dangerous substances less dangerous, which makes the environment for technicians and researchers safer. These technologies also better public health outcomes by lowering exposure to air, water, and soil toxins by minimizing environmental pollutants and hazardous waste [26].

Glycan extraction from biological materials takes a lot less time and energy when using microwave-assisted extraction (MAE). According to studies, MAE can cut extraction times from hours to minutes, saving a significant amount of energy. When compared to traditional extraction techniques, MAE frequently uses less solvent, which lowers the amount of hazardous waste produced and the expense of disposing of the used solvents. Supercritical CO2 is a generally available, non-flammable mobile phase used in supercritical fluid chromatography (SFC). By doing this, the negative effects of using organic solvents on the environment are lessened. Compared to conventional liquid chromatography, SFC offers high-resolution separation with less energy usage, making it a more environmentally friendly choice for glycan analysis. ILs are less prone to evaporate and pollute the environment because of their low volatility. They may be adjusted to minimize environmental concerns and be used in a variety of separation procedures. Reusing and recycling a large number of ILs can cut down on the amount of trash generated overall and the requirement for new solvents in successive separation cycles [27].

Using environmentally friendly technologies in glycan separation has major financial benefits in addition to positive environmental effects. For assessing the viability and effectiveness of these green technologies, it is crucial to comprehend the economic factors and apply sustainability measures. The economic advantages, cost reductions, and important sustainability indicators related to glycan separation are covered in detail in this section. Green technologies generally employ solvents and reagents that are more affordable and environmentally friendly. For example, using water or supercritical CO2 in place of conventional organic solvents can drastically reduce expenses. Reducing the production of hazardous waste lowers the costs associated with disposal and environmental regulation compliance. Energy-efficient procedures, such extraction with microwave assistance and separation with ultrasound assistance, cut utility expenses by using less energy and requiring less time to complete. Green technologies frequently simplify separation procedures, cutting down on the number of steps and labor expenses involved. Faster throughput and more productivity may result from this. It is easier to comply with regulations and less likely to incur fines and penalties when using safer, environmentally friendly chemicals instead of dangerous ones. Businesses that use green technologies could be able to obtain sustainable certifications, which would increase their marketability and possibly allow them to charge higher rates for their goods. The market for things made responsibly is expanding. Businesses can take advantage of this need to set themselves apart in the market by utilizing green glycan separation methods. Adopting eco-friendly procedures can improve a business' standing and brand value, drawing in eco-aware investors and clients [26].

How well reactants are transformed into desired products is how atom economy gauges the effectiveness of a chemical process. A higher atom economy means less waste and by-products. Efficient derivatization procedures that include all reactants into the final analyte increase atom economy and minimize waste in glycan separation. The quantity of waste generated per unit of product is measured by the E-factor. A procedure that is more ecologically friendly has a lower E-factor. Green glycan separation technologies use less solvent and produce less waste in an effort to reduce the E-factor. Energy intensity measures how much energy is used for each unit of a product. A procedure that uses less energy is said to be more energy-efficient. Glycan separation procedures use less energy when energy-saving methods like MAE are used. The entire amount of freshwater required to produce a product is measured by its "water footprint." Reduced water use is a sign of a more environmentally friendly procedure. Water is a solvent that is frequently used in green separation processes; this reduces the overall water footprint because water can be recycled and reused. The evaluation of solvents and reagents' toxicity and biodegradability guarantees that the chemicals employed do not linger in the environment or present serious health hazards. An indicator of the sustainability of the materials used in the process is the ratio of renewable to non-renewable feedstocks. Supercritical CO2 is a safer and less expensive alternative to conventional organic solvents. Cost effectiveness is further increased by the capacity to recycle CO2. This process uses less energy and produces less waste while having a high atom economy and low E-factor. ILs can be more expensive at first, but over time, their reduced waste production and ability to be recycled can make up for this. Low toxicity and biodegradability measurements are often attributed to ILs' great thermal stability and low volatility [27].

6.4 Green Solvent and Solvent-Free Techniques

The discovery and use of ecologically friendly solvents and solvent-free glycan separation methods are results of the quest for green chemistry. The advantages and uses of ILs, DESs, and solvent-free extraction techniques are discussed in this section with special attention to glycan separation.

6.4.1 Green Solvent Techniques

6.4.1.1 Ionic Liquids

Salts that are liquid and at or close to room temperature are known as ionic liquids or ILs. They are made up of an inorganic or organic anion and an organic cation. Because of their special qualities—like their tunable solubility, great thermal stability, and minimal vapor pressure—ILs have attracted interest. Because ILs have extremely low vapor pressures, there is less chance of exposure to volatile organic chemicals and air pollution. ILs are appropriate for a variety of high-temperature operations because they can tolerate high temperatures without breaking down. By altering the cation or anion, it is possible to customize the physical and chemical characteristics of ILs for certain uses. Glycans can be selectively extracted by ILs from intricate biological matrices, increasing the yield and efficiency of separation. Because of their distinct solvation characteristics, ILs are employed as stationary phases in chromatography to improve the separation of glycan isomers [28].

6.4.1.2 Deep Eutectic Solvents

A hydrogen bond donor and a hydrogen bond acceptor are combined to create DESs, which are eutectic mixtures with melting points that are much lower than those of the constituent parts. DESs and ILs both have many green qualities; however, DES production is frequently simpler and less expensive. Comparing DESs to various ILs and traditional solvents, they are typically less hazardous and more biodegradable. Simple, affordable ingredients like urea and choline chloride can be used to create DESs. DESs are adaptable for a variety of applications because they can dissolve a broad spectrum of compounds. DESs offer a green substitute for conventional solvents by effectively dissolving glycans and facilitating their extraction from natural sources. Enzymes employed in glycan modification and analysis can be stabilized by DESs, improving the efficiency and yield of the reaction [29].

6.4.2 Solvent-Free Extraction Method

Solvent-free extraction techniques seek to completely do away with the use of solvents, minimizing their negative effects on the environment and enhancing safety. These techniques make use of physical processes to accomplish effective separation and extraction. Target chemicals are released and cell disruption is caused by direct heating of the sample utilizing microwave energy in MAE. Compared to traditional procedures, MAE shortens extraction times, uses less energy, and frequently requires no or very little solvent. By using ultrasonic waves to induce cavitation bubbles in the sample, ultrasound-assisted extraction (UAE) releases glycans when the bubbles rupture and disrupt cell walls. UAE preserves the integrity of heat-sensitive glycans and improves extraction efficiency at lower temperatures. Supercritical CO_2 is used in supercritical fluid extraction as a solvent to extract various chemicals. In its supercritical form, CO_2 exhibits solvation characteristics similar to liquids and diffusivity similar to gases. Because CO_2 is non-toxic, non-flammable, and recyclable, SFE is environmentally benign. Furthermore, it enables exact control over the circumstances of extraction [30].

6.4.3 Industrial Applications of Green Solvent Techniques

In order to ensure high purity and safety standards, the pharmaceutical industry uses solvent-free and green solvent extraction and analysis procedures for glycans in medication development and quality control. Glycoproteins and other biopharmaceuticals are purified using DESs and ILs, which increase yield and activity. Glycans are separated into useful food ingredients, including prebiotics, using green extraction procedures, which guarantee the sustainability and safety of the final product.

DESs are employed to minimize environmental effect while enhancing health advantages by extracting bioactive glycans from plant sources for application in nutraceuticals. In biotechnological applications, DESs and ILs facilitate enzymatic glycan changes, which increase process sustainability and efficiency. The examination of intricate glycan structures is improved by solvent-free techniques, which yield precise and trustworthy data for study and innovation [29].

6.5 Membrane Technologies for Glycan Separation

Membrane technologies have shown to be effective instruments for glycan separation and purification. High efficiency, scalability, and the possibility of integrating with green chemistry principles are only a few benefits of these techniques. This section examines reverse osmosis (RO) and nanofiltration, membrane filtration and ultrafiltration (UF), as well as the advancement of green membrane materials and their uses in glycan separation.

6.5.1 Membrane Filtration and Ultrafiltration

Semi-permeable membranes are used in membrane filtration to separate compounds according to their sizes and other characteristics. This method is frequently applied in the bioprocessing sector to concentrate and purify biomolecules, such as glycans. A solution is passed across a membrane that has pores of a particular size during the membrane filtration process, allowing smaller molecules to pass through while holding larger molecules in place. With pore diameters typically ranging from 0.1 to 10 μm, it can be utilized for microfiltration, which is effective in eliminating larger particles and bacteria but not smaller glycans. Glycan-containing solutions are clarified by membrane filtration, which also removes other particulate matter and cell debris, readying the sample for additional purification. It can also function as the first stage of purification, concentrating glycans by the elimination of bigger impurities [31].

Smaller molecules, such as proteins and glycans, can be separated using UF, a subtype of membrane filtration with pore diameters typically between 1 and 100 nm. Moderate pressures are used in UF to push the fluid through a membrane that keeps larger molecules in place while allowing smaller molecules and ions to pass. The membrane's molecular weight cutoff serves as the main basis for separation. By eliminating water and low-molecular-weight contaminants, UF concentrates glycans to a higher concentration for further processing or analysis. Salts and tiny compounds may be effectively removed with UF, which also makes buffer exchange and glycan preparation easier for use in subsequent processes [32].

6.5.2 Nanofiltration and Reverse Osmosis

With typical pore diameters of 1 to 10 nm, nanofiltration (NF) fills the void left by UF and RO. It is appropriate for some glycan separations because it can separate molecules according to their size and charge. Smaller pore diameters in nanofiltration membranes allow smaller molecules and monovalent ions to flow through while trapping bigger solutes and multivalent ions inside. In order to balance efficiency and energy consumption, NF runs at pressures that are higher than UF but lower than RO. By eliminating impurities based on size and charge, NF can purify glycans and increase the preparation's purity. Glycan solutions are desalinated using NF, which efficiently eliminates salts and other minute contaminants [31].

The smallest hole diameters used in membrane technologies are in RO, usually less than 1 nm, which only permits water and extremely small molecules to pass through. Because of this, RO is very useful in separation operations that call for high purity. High pressure is applied to a solution during RO, pushing water molecules through a semipermeable membrane while keeping bigger solutes like salts and organic molecules in place. This method may remove tiny chemical compounds and dissolved salts almost entirely. In order to produce high-purity glycans, RO is utilized for the last stage of glycan solution purification, guaranteeing the elimination of all salts and minute impurities. RO is an important method for creating extremely pure glycan samples because it can simultaneously concentrate and purify glycan solutions [32].

6.5.3 Green Membrane Materials and Applications

Improving the sustainability of membrane technologies requires the development of green membrane materials. These materials seek to enhance the lifetime and effectiveness of membrane-based processes while lessening their negative effects on the environment. Because of their environmental friendliness and biodegradability, biodegradable polymers like polylactic acid and polyhydroxyalkanoates are being investigated as potential membrane materials. These materials are employed in the production of glycan separation membranes, which lessen the separation process' environmental impact. Membranes made with recycling and reuse in mind assist cut down on waste production and membrane replacement expenses.

Many glycan separation procedures use reusable membranes, which provide both long-term sustainability and economic viability. The overall sustainability of membrane technologies is enhanced by the synthesis of membrane materials employing green chemistry concepts, such as solvent-free processes and the utilization of renewable feedstocks. By producing membranes for glycan separation using these green synthesis approaches, environmental objectives are met and process sustainability is enhanced [33].

6.6 Adsorption and Chromatographic Methods

The two main methods for separating glycans are chromatography and adsorption. In order to minimize environmental effect while maintaining high efficiency and selectivity, recent improvements have concentrated on developing environmentally friendly chromatographic procedures including green adsorbents and stationary phases. These developments and their uses in glycan separation are covered in this section.

6.6.1 Green Adsorbents

The interaction of the adsorbent material with the target molecules is the foundation of adsorption techniques. Green adsorbents are environmentally beneficial substitutes for conventional materials because they are made from renewable, biodegradable, or sustainable resources. Chitosan, which is derived from chitin, is a biodegradable and biocompatible polymer that has a high capacity for adsorbing glycans because of its amino groups, which can bind electrostatically and through hydrogen bonds with glycan molecules. Glycans are adsorbed onto modified cellulose materials, such as carboxymethyl cellulose. They are appealing for environmentally friendly applications because of their biodegradability and availability from renewable sources. Because of its high surface area and porous structure, biochar, which is created through the pyrolysis of biomass, is a highly effective adsorbent for glycans. Moreover, it emits no carbon, promoting environmental sustainability. Glycans can be extracted from aqueous solutions using biochar, offering an environmentally friendly way to concentrate and purify glycans. The selectivity of silica materials for glycan adsorption is improved when they are functionalized with organic groups (such as amino or thiol). These materials have customizable surface characteristics together with good mechanical stability. Green chemistry techniques, such sol-gel procedures in aqueous media, can be used to synthesize silica-based adsorbents with less hazardous solvent usage [34].

6.6.2 Green Stationary Phases

Stationary phases are essential in chromatography because they allow glycans to be separated according to how they interact with the stationary material. Green stationary phases are made with efficiency and environmental friendliness in mind. In SEC, stationary phases based on agarose and dextran are frequently employed to separate glycans. These biopolymers come from renewable resources and are biodegradable. These stages offer exceptional resolution and repeatability when it comes to size-based glycan separation. High selectivity and durability are offered by hybrid materials that combine silica with organic functional groups (such as amine or phenyl). Green synthesis techniques can be employed to improve the environmental profile of these materials. These hybrids are utilized for the effective separation of glycans in a variety of chromatographic procedures, including reversed-phase and hydrophilic interaction liquid chromatography [35].

6.6.3 High-Performance Liquid Chromatography

One flexible method for glycan separation is high-performance liquid chromatography (HPLC). HPLC is now more environmentally friendly thanks to the development of eco-friendly solvents and procedures. Toxic organic solvents are used less frequently in HPLC when water and ethanol are used as mobile phases. These solvents present less of a risk to laboratory workers' health and the environment. High resolution and sensitivity are provided by green solvent systems, which are utilized in both normal-phase and reversed-phase HPLC for glycan analysis. Supercritical CO_2 is the main mobile phase in SFC, and it is occasionally mixed with a tiny quantity of organic modifier. Because CO_2 is widely available, non-flammable, and non-toxic, SFC is a more environmentally friendly method of liquid chromatography than the conventional method. Glycan separation uses SFC, which has a high efficiency and quick analytical times. It works especially well with non-polar glycans [36].

6.6.4 Thin-Layer Chromatography

Thin-layer chromatography (TLC) is an easy and affordable chromatographic technique. Sustainable changes have increased its sustainability. It is eco-friendly to use silica gel and cellulose as stationary phases in TLC. These components come from natural sources and are biodegradable. Glycans are analyzed qualitatively using TLC with green stationary phases, which offers a quick and easy way to keep an eye on the separation processes. The amount of dangerous organic solvents used in TLC is reduced by using combinations of water and ethanol as eluents. Both safety and the environment are enhanced by this method. Polar glycan separation makes use of green eluents, which provide high resolution and repeatability [36].

6.6.5 Capillary Electrophoresis

CE is an effective method for analyzing glycans. Using eco-friendly reagents and reducing waste are the main goals of green approaches in CE. By using aqueous buffers instead of organic solvents, CE is a more ecologically responsible technique. High resolution and quick analysis are achieved by the separation of glycans using environmentally acceptable buffers. Microchip CE systems minimize waste production and reagent use by using very small quantities of samples and reagents. High-throughput glycan analysis using Microchip CE, which provides quick, is an effective separation with no environmental impact [34].

6.7 Biocatalytic and Enzymatic Approaches

Glycan modification and separation have undergone a revolutionary shift thanks to biocatalysis and enzymatic methods, which offer great specificity, efficiency, and environmental friendliness. These methods lessen the requirement for harsh chemicals and harsh environments by using enzymes to alter glycans and catalyze reactions under mild settings. This chapter covers the development and implementation of these techniques in biotechnology, as well as enzyme-based glycan modification and separation, biocatalytic procedures for environmentally friendly glycan purification, and more.

6.7.1 Enzyme-Based Glycan Modification and Separation

Enzymes are unique biocatalysts with the capacity to identify and change specific glycan structures. They are very useful tools for glycan modification and separation because of their specificity. The hydrolysis of glycosidic linkages in glycans by glycosidases releases monosaccharides and smaller oligosaccharides. These enzymes have the ability to break particular links within complex glycan structures in a selective manner. In order to investigate the structures and functions of glycans, glycosidases are utilized in the deglycosylation of glycoproteins. In order to simplify glycan mixtures for additional analysis, they are also employed in the elimination of certain glycan residues. Glycosyltransferases catalyze the formation of glycosidic linkages by moving sugar moieties from activated donor molecules to certain acceptor molecules. The production and remodeling of glycans depend on this process. These enzymes are employed to create particular glycans with desirable structural configurations. Sialyltransferases add sialic acid residues to the termini of glycan chains, which is crucial for the biological activity and stability of glycans. By adding sialic acid residues, which can improve circulation time and reduce immunogenicity, sialyltransferases are used to enhance the stability and therapeutic efficacy of glycoprotein-based drugs. Using different glycosyltransferases allows researchers to build complex glycan structures with high precision, facilitating the study of glycan functions and interactions [26].

6.7.2 Biocatalytic Processes for Eco-Conscious Glycan Purification

Enzyme specificity and mild reaction conditions are used by biocatalytic techniques to produce environmentally friendly glycan purification. The usage of dangerous chemicals and energy-intensive settings is reduced by these procedures. Using glycosidases, enzymatic deglycosylation entails removing specific glycan residues from glycoproteins or complicated glycan mixes. Targeting particular glycosidic bonds is possible with this very specialized technique. Extreme temperatures and harsh chemicals are less necessary when enzymatic deglycosylation takes place in benign environments like room temperature and neutral pH. As a result, the procedure is energy and environmentally efficient. During enzymatic synthesis and coupling, certain sugar residues are added to acceptor molecules by glycosyltransferases, resulting in the formation of new glycosidic linkages. This technique makes it possible to precisely create glycan structures. Organic solvents and dangerous chemicals are used less frequently in enzymatic synthesis and coupling reactions because they typically work in moderate environments and employ aqueous media. This encourages sustainability and is consistent with the ideas of green chemistry. Enzymes are used in biocatalytic glycan separation to change glycan structures in a selective manner that makes them amenable to separation methods like chromatography. To simplify the glycan mixture for simpler separation, various glycosidases, for instance, can be employed to cleave particular bonds. The separation process can be completed in mild circumstances with the use of biocatalysts, which lowers energy consumption and waste product generation. This method improves glycan purification's overall sustainability [35].

6.7.3 Advances and Applications in Biotechnology

The fields of glycan research and industry have found new uses for biocatalytic and enzymatic techniques thanks to recent developments in biotechnology. Enzymes with the required specificities and activity can now be quickly identified and optimized thanks to high-throughput screening technology. This has made it easier to find new enzymes for the modification and separation of glycans. These developments enable the creation of specialized enzyme cocktails for certain glycan purification tasks, improving glycan processing efficiency and selectivity. Enzymes can be more stable and reusable when immobilized on solid substrates, which increases the efficiency and scalability of biocatalytic processes. For the purpose of immobilizing enzymes, methods like entrapment, adsorption, and covalent bonding are employed. Continuous-flow bioreactors are used to produce and purify glycans on a wide scale using immobilized enzyme systems. These technologies reduce labor and operating expenses by providing constant performance and being able to be integrated into automated processes. Enzyme characteristics, including substrate selectivity, activity, and stability, can be changed through the use of directed evolution techniques and enzyme engineering. These techniques entail introducing mutations and choosing versions that perform better. Novel glycan structures and enhanced glycan modification and separation processes are made possible by engineered enzymes. These enzymes are designed to fulfil particular industrial requirements, boosting the synthesis of medicines and diagnostics based on glycans [31].

6.8 Electrochemical and Electrophoretic Techniques

Because of their great sensitivity, specificity, and potential for environmentally benign applications, electrochemical and electrophoretic techniques are essential in the separation and study of glycans. This chapter explores the fundamentals of electrochemical glycan separation, green glycan analysis using electrophoretic techniques, and upcoming technologies that will influence this subject going forward.

6.8.1 Electrochemical Glycan Separation Principles

Glycan molecules' special redox characteristics are used by electrochemical techniques to separate and analyze them. These methods have benefits like quick analysis, less sample preparation, and the use of safe, environmentally friendly reagents. Measuring the current produced by the oxidation or reduction of glycan molecules at an electrode surface is known as electrochemical detection. This current varies in direct proportion to the analyte's concentration. To increase the sensitivity of their detection, glycans can be derivatized with electroactive labels. To identify and quantify glycan species, this method is employed in conjunction with capillary electrophoretic separation or chromatographic separation. Electrochemical biosensors preferentially bind glycans by using biological recognition components, such as glycosidases or lectins, that have been immobilized on electrode surfaces. There is a detectable electrochemical signal produced by the binding event. These biosensors are used to track glycan concentrations and interactions in complicated biological samples in real-time. They are appropriate for glycomics research and clinical diagnostics because of their excellent specificity and sensitivity. Glycans are separated in electrochemical CE according to their charge-to-mass ratio in the presence of an electric field. One extremely sensitive method of quantification is capillary outlet electrochemical detection. With this method, glycan mixtures can be separated quickly and effectively, yielding comprehensive profiles of the compositions and structures of the glycans [37].

6.8.2 Electrophoretic Methods for Green Glycan Analysis

Using an electric field, electrophoretic methods separate glycans according to their charge and size. Because these techniques require less chemicals and organic solvents, they are naturally green. By directing a strong electric field across a small capillary that is filled with an electrolyte, CE separates glycans. Glycans travel at varying speeds according on their charge and size. CE produces less waste because it uses extremely small amounts of reagents and samples. It does not use dangerous organic solvents because it works in aqueous media. In investigations on glycomics, biotherapeutics, and pharmaceuticals, CE is frequently utilized for the study of glycan profiles. It offers quick analytical times and separation with high resolution. Glycans are separated using a gel matrix and an electric field in gel electrophoresis. Smaller molecules can go through the gel more quickly than larger ones because it functions as a molecular sieve. In gel electrophoresis, non-toxic reagents and buffers are usually used. Developments in electrophoresis equipment and gel materials have significantly reduced environmental effect. This method is frequently applied to the qualitative investigation of glycans, especially in the characterization and quality control of glycoproteins. Glycans are separated *via* isoelectric focusing (IEF) according to their isoelectric points (pI) along a pH gradient. The point on the gradient where glycans have no net charge is where they migrate. IEF reduces chemical waste by using small amounts of buffers and ampholytes. By using a regulated pH environment, the approach reduces the amount of harsh chemicals used. Glycans and glycoproteins can be separated at high resolution using IEF, which helps identify isomeric and structurally identical glycans [38].

6.8.3 Emerging Technologies and Future Directions

Emerging technologies in the field of electrochemical and electrophoretic glycan analysis provide promise for improving sustainability, sensitivity, and efficiency. The field is changing quickly. Microfluidic devices allow for high-throughput and miniaturized glycan research by combining electrochemical and electrophoretic procedures on a single chip. By using relatively small volumes of chemicals and samples, these machines drastically cut down on waste. Additionally, they provide the possibility of real-time and on-site analysis, which eliminates the requirement for a large laboratory infrastructure. Glycan analysis is anticipated to undergo a revolution with the introduction of more advanced microfluidic platforms including integrated detection systems, thereby rendering it more ecologically benign and widely accessible. Utilizing nanomaterials for glycan detection increases the sensitivity and specificity of electrochemical sensors. Examples of these materials are graphene, carbon nanotubes, and metal nanoparticles. Using less material and energy, nanomaterials can enhance sensor performance. The necessity for thorough sample preparation and purification may be lessened by their implementation into biosensors. More reliable and effective glycan analysis sensors will be produced with ongoing advancements in nanomaterial synthesis and functionalization, which will have applications in environmental monitoring, biotechnology, and healthcare. Greener alternatives to conventional gel materials are provided by the development of sustainable and biodegradable electrophoretic media, such as natural polysaccharides and synthetic biodegradable polymers. These medium minimize the amount of non-biodegradable waste generated during gel electrophoresis, thereby lessening its environmental impact.

They also improve the sustainability and safety of electrophoretic methods. Ongoing research endeavors to enhance the efficacy and durability of these eco-friendly materials, with the goal of attaining parity or even greater separation efficiency in contrast to traditional materials [39].

6.9 Integrated Green Technologies

To achieve greater efficiency, sustainability, and performance, several environmentally friendly methods are included in the integration of green technologies for glycan separation. This chapter examines integrated and hybrid methods for separating glycans, the benefits of integrating green methods, and industrial implementation optimization methodologies.

6.9.1 Hybrid and Integrated Approaches to Glycan Separation

To improve the efficiency and sustainability of the process, hybrid and integrated approaches to glycan separation incorporate a variety of environmentally friendly methodologies. By utilizing the advantages of each unique technique, these methods produce separation protocols that are both more effective and ecologically benign. Combining chromatography and electrophoresis makes use of both processes' great resolution and quickness. Methods like capillary electrochromatography combine liquid chromatography with capillary electrophoresis (CE). These hybrids provide thorough structural information and increase throughput by being utilized for high-resolution separation and quick analysis of intricate glycan combinations. In glycan analysis, integrating electrochemical detection with CE improves selectivity and sensitivity. A sensitive quantification technique is offered by electrochemical detection at the capillary exit. Glycans in biological samples may be precisely quantified with the use of these technologies, which makes glycomics and biotherapeutics research easier. On a single chip, microfluidic devices can combine several separation methods, including electrophoresis, chromatography, and electrochemical detection. Glycan analysis can be done automatically and continuously thanks to this integration. In clinical diagnosis and research, microfluidic systems are utilized for high-throughput screening, real-time monitoring, and comprehensive glycan profiling [40].

6.9.2 Synergistic Effects of Combined Green Techniques

When green techniques are combined, there might be synergistic effects that outweigh the contributions of each strategy alone. Improved separation performance, decreased environmental impact, and increased efficiency are the results of these synergistic effects.

Higher separation efficiency and resolution can be attained by combining CE with HPLC. Glycans are first separated by HPLC using hydrophobicity as a basis, and then they are further separated by CE using charge-to-mass ratio. This combination makes it possible to resolve complicated glycan mixtures that would be difficult to separate with a single method, leading to an analysis that is more precise and thorough. The usage of dangerous organic solvents is decreased when green solvents, such as DESs or ILs, are combined with electrophoretic methods. Because these green solvents are biodegradable and less harmful, their impact on the environment is reduced. By lowering the production of hazardous waste and raising the general level of safety throughout the separation process, this integration encourages sustainability in glycan analysis. The sensitivity and selectivity of glycan analysis are increased when electrochemical detection is combined with chromatography or electrophoresis. Glycan molecules may be precisely detected at low concentrations using electrochemical sensors. This method makes it possible to identify and measure trace glycan species in intricate biological samples, which helps with research on the identification of biomarkers and the diagnosis of disease [40].

6.10 Future Trends and Prospects

Glycan separation is positioned for major progress, propelled by new green technologies, creative discoveries, and the quest of environmentally friendly methods. This chapter examines the prospects and future trends in glycan research, emphasizing the opportunities and difficulties that lie ahead as well as the potential for new technologies and breakthroughs. New green technologies are revolutionizing glycan research by offering more environmentally friendly, productive, and sustainable glycan separation and analysis techniques. The goal of these technologies is to maximize analytical performance while minimizing the impact on the environment. Adsorption and chromatography are two separation methods that employ nanomaterials to increase the sensitivity and selectivity of glycan analysis. For effective separation and purification, magnetic nanoparticles can be functionalized to preferentially trap particular glycan molecules.

High-throughput glycan analysis is performed in research and clinical diagnostics using microfluidics and lab-on-a-chip systems, which offer quick, accurate results with no environmental impact. To further improve sustainability, microfluidic devices can integrate a variety of green technologies, including bio-based materials and green solvents. Green solvent systems, which offer efficient separation with less of an adverse influence on the environment, are being utilized more and more in chromatographic and extraction procedures for glycan analysis. Green solvent chemistry developments keep enhancing performance and broadening the scope of applications.

Advances in glycan separation are opening doors for future discoveries that could completely transform the industry. These developments should improve the effectiveness, precision, and long-term viability of glycan research. To improve data analysis, optimize experimental conditions, and create predictive models for glycan separation, artificial intelligence (AI) and machine learning are employed. These instruments have the potential to greatly speed up research and development, which will hasten the identification of novel glycan-based biomarkers and treatments. For glycan analysis, 3D printed separation columns, membranes, and microfluidic devices are being developed; these materials offer scalable and customizable options. Rapid prototyping and production are made possible by advanced manufacturing, which supports the creation of cutting-edge separation technologies. Glycan separation involves the use of biomimetic membranes, enzyme-mimicking catalysts, and bio-inspired adsorption materials. These technologies are appealing for a variety of glycomics and biotechnology applications due to their great selectivity, efficiency, and sustainability.

There are several potential and problems in the development and application of green technology for glycan separation. To advance the area and realize the full potential of sustainable glycan research, these obstacles must be overcome. It can be difficult to scale up green technology from the laboratory to the industrial scale because of variations in process dynamics, financial constraints, and legal restrictions. Industrial adoption can be facilitated by creating green technologies that are affordable, scalable, and validated using strong optimization and validation techniques. This process can be sped up by cooperation between regulatory agencies, business, and academia. When compared to conventional techniques, the initial cost of introducing green technologies may be higher. Achieving economic sustainability is essential for broad adoption, especially in environments with limited resources. Green technologies can become more economically viable and less expensive with the help of developments in green chemistry, materials science, and process engineering. Cost-benefit analysis and life cycle assessment can be used to show long-term economic benefits and pinpoint areas where costs can be cut. It can be challenging to achieve standardization across various green technologies and to ensure compliance with regulatory standards. Regulations need to change to make room for fresh, creative approaches. The acceptance and application of green technologies can be aided by the creation of uniform and transparent regulatory frameworks. Addressing these issues requires collaboration with standardization organizations and regulatory organizations.

6.11 Conclusion

We have examined the state-of-the-art environmentally responsible glycan separation technologies in this chapter, highlighting their significance and possible influence on the glycomics research. Green solvents, membrane technologies, adsorption and chromatographic techniques, biocatalytic and enzymatic procedures, electrochemical and electrophoretic methods, and other green technologies were discovered and discussed. Every technology has its own advantages and solutions for particular glycan separation problems. In addition to lowering energy use and hazardous waste production, green technologies also provide financial benefits by using sustainable and affordable techniques. Even if green technologies have a lot of potential, they also have drawbacks like regulatory compliance, economic feasibility, and scalability. It will take constant innovation, teamwork, and supporting policy to overcome these obstacles. The efficiency and efficacy of glycan separation can be increased by using integrated and hybrid strategies that incorporate several green technologies. In order to optimize processes and achieve industrial implementation, these synergistic effects are essential.

Green technologies enhance the glycan separation processes' selectivity, sensitivity, and throughput by utilizing improved materials, microfluidics, and AI-driven optimization. This allows for more precise and thorough glycomics research. Green technology use in industrial settings can result in more environmentally friendly manufacturing operations, which will lessen the impact of glycan analysis and production. This complies with legal standards as well as global sustainability goals. Green technologies are more economically viable due to their cost-effectiveness as well as their capacity to lower waste and energy consumption. Because of this, they are appealing for widespread adoption, especially in settings with limited resources. Clear legal frameworks and the standardization of green technology will help them become more widely accepted and used, which will improve glycan analysis' consistency and dependability. Going forward, glycan separation will be shaped by ongoing innovation as well as the incorporation of cutting-edge green technology. New developments in biomimetic methods and nanotechnology will provide fresh ways to glycan separation, improving productivity and sustainability. Research and development will move more quickly with the use of AI and machine learning to optimize data analysis and separation procedures, producing better results and discoveries more quickly. The development and application of green technology will be accelerated by increased cooperation between academic institutions, business, and regulatory agencies, guaranteeing that they satisfy the demands of diverse stakeholders. A culture of sustainability in glycomics research will be fostered by raising awareness and providing education on the advantages and potential of green technologies. This will facilitate their adoption and implementation.

The field of glycan separation can make great strides toward the development of glycomics and environmental sustainability by embracing these future directions. This thorough investigation of environmentally friendly glycan separation technology emphasizes how crucial innovation and sustainability are to the advancement of glycomics research. It is possible for the discipline to achieve higher efficiency, environmental stewardship, and scientific discoveries by embracing green concepts and cutting-edge technologies.

AI disclosure statement: The author confirms that no artificial intelligence (AI)-generated content or AI-assisted writing tools were used in the preparation of this chapter (including summaries and the abstract). All text, analysis, and conclusions are the result of the author's original work.

References

1. Daramola, O., Gutierrez-Reyes, C.D., Wang, J., Nwaiwu, J., Onigbinde, S., Fowowe, M., *et al.*, Isomeric separation of native N-glycans using nano zwitterionic-hydrophilic interaction liquid chromatography column. *J. Chromatogr. A*, 1705, 464198, 2023.
2. Ma, X., Li, M., Tong, P., Zhao, C., Li, J., Xu, G., A strategy for construction of highly sensitive glycosyl imprinted electrochemical sensor based on sandwich-like multiple signal enhancement and determination of neural cell adhesion molecule. *Biosens. Bioelectron.*, 156, 112150, 2020.
3. Wang, C., Gao, X., Gong, G., Man, L., Wei, Q., Lan, Y., *et al.*, A versatile strategy for high- resolution separation of reducing glycan mixtures as hydrazones

by two-dimensional high-performance liquid chromatography. *J. Chromatogr. A*, 1685, 463599, 2022.

4. Qing, G., Yan, J., He, X., Li, X., Liang, X., Recent advances in hydrophilic interaction liquid interaction chromatography materials for glycopeptide enrichment and glycan separation. *TrAC, Trends Anal. Chem.*, 124, 115570, 2020.
5. Xie, C., Wu, Q., Zhang, S., Wang, C., Gao, W., Yu, J., *et al.*, Improving glycan isomeric separation *via* metal ion incorporation for drift tube ion mobility-mass spectrometry. *Talanta*, 211, 120719, 2020.
6. Kurz, S., Sheikh, M.O., Lu, S., Wells, L., Tiemeyer, M., Separation and identification of permethylated glycan isomers by reversed phase nanoLC-NSI-MSn. *Mol. Cell. Proteomics*, 20, 100045, 2021.
7. Arora, K., Sherilraj, P.M., Abutwaibe, K.A., Dhruw, B., Mudavath, S.L., Exploring glycans as vital biological macromolecules: A comprehensive review of advancements in biomedical frontiers. *Int. J. Biol. Macromol.*, 268, 1, 131511, May 2024.
8. Yao, G., Ke, W., Xia, B., Gao, Z., Nanopore-based glycan sequencing: state of the art and future prospects. *Chem. Sci.*, 15, 6229, 2024.
9. Adamczyk, B., Tharmalingam-Jaikaran, T., Schomberg, M., Szekrényes, A., Kelly, R.M., Karlsson, N.G., *et al.*, Comparison of separation techniques for the elucidation of IgG N- glycans pooled from healthy mammalian species. *Carbohydr. Res.*, 389, 174–85, 2014.
10. Hecht, E.S., Obiorah, E.C., Liu, X., Morrison, L., Shion, H., Lauber, M., Microflow size exclusion chromatography to preserve micromolar affinity complexes and achieve subunit separations for native state mass spectrometry. *J. Chromatogr. A*, 1685, 463638, 2022.
11. Zhang, Y. and Li, J., Separation of saccharides by ion-exchange chromatography, in: *Ion-Exchange Chromatography and Related Techniques*, pp. 371–85, Elsevier, 2024.
12. Farsang, R., Jarvas, G., Guttman, A., Purification free N-glycan analysis by capillary zone electrophoresis: hunt for the lost glycans. *J. Pharm. Biomed. Anal.*, 238, 115812, 2024.
13. Yamamoto, S., Ueda, M., Kasai, M., Ueda, Y., Kinoshita, M., Suzuki, S., A fast and convenient solid phase preparation method for releasing N-glycans from glycoproteins using trypsin-and peptide-N-glycosidase F (PNGase F)-impregnated polyacrylamide gels fabricated in a pipette tip. *J. Pharm. Biomed. Anal.*, 179, 112995, 2020.
14. Harvey, D.J., Royle, L., Radcliffe, C.M., Rudd, P.M., Dwek, R.A., Structural and quantitative analysis of N-linked glycans by matrix-assisted laser desorption ionization and negative ion nanospray mass spectrometry. *Anal. Biochem.*, 376, 44–60, 2008.
15. Zhang, Y.-Y., Zhang, S.-Y., Hu, Z.-X., Voglmaier, J., Liu, L., Galan, M.C., *et al.*, High sensitivity profiling of N-glycans from mouse serum using fluorescent imidazolium tags by HILIC electrospray ionisation spectrometry. *Carbohydr. Polym.*, 343, 122449, 1 November 2024.

16. Meudt, M., Baumeister, J., Mizaikoff, B., Ebert, S., Rosenau, F., Blech, M., *et al.*, Comprehensive analysis and characterization of glycan pairing in therapeutic antibodies and Fc-containing biotherapeutics: Addressing current limitations and implications for N-glycan impact. *Eur. J. Pharm. Biopharm.*, 200, 114325, 2024.
17. Borah, D., Mishra, V., Debnath, R., Ghosh, K., Gogoi, D., Rout, J., *et al.*, Facile green synthesis of highly stable, water dispersible carbohydrate conjugated Ag, Au and Ag-Au biocompatible nanoparticles: catalytic and antimicrobial activity. *Mater. Today Commun.*, 37, 107096, 2023.
18. Linssen, M.D.M.E., Lin, Y.-T., van den Wildenberg, S.A.H., Tholen, M.M.E., de Jong, A.M., Prins, M.W.J., Oriented Antibody Coupling to an Antifouling Polymer Using Glycan Remodeling for Biosensing by Particle Motion. *Bioconjugate Chem.*, 35, 7, 996–1006, 2024.
19. Verma, C., Chauhan, D.S., Aslam, R., Banerjee, P., Aslam, J., Quadri, T.W., *et al.*, Principles and theories of green chemistry for corrosion science and engineering: design and application. *Green Chem.*, 26, 4270–357, 2024.
20. Hariprasad, B.S., Priya, R.L., Babu, S.G., Recent advances in the synthesis, catalytic, and adsorption applications of carbohydrate polymers. *Eur. Polym. J.*, 214, 113178, 24 June 2024.
21. Dalastra, C., Klanovicz, N., Kubeneck, S., Stefanski, F.S., Argenta, D.F., Rauber, G.S., *et al.*, Carbohydrate-based economy: perspectives and challenges, in: *Polysaccharide-degrading biocatalysts*, pp. 409–34, 2023.
22. You, Y., Kong, H., Li, C., Gu, Z., Ban, X., Li, Z., Carbohydrate binding modules: Compact yet potent accessories in the specific substrate binding and performance evolution of carbohydrate-active enzymes. *Biotechnol. Adv.*, 73, 108365, July–August 2024.
23. Taoufik, A., Oudghiri, K., Taourirte, M., Agunaou, M., Bahsis, L., Recent advances in carbohydrate polymers as a catalyst for organic synthesis: An update. *Synth. Commun.*, 54, 11, 853–880, 2024. https://doi.org/10.1080/00397911.2024.2337090.
24. Liu, Y., Yang, T., Rong, J., Yuan, J., Man, L., Wei, M., *et al.*, Integrated analysis of natural glycans using a versatile pyrazolone-type heterobifunctional tag ANPMP. *Carbohydr. Polym.*, 327, 121617, 2024.
25. Upadhyaya, K. and Dubbu, S., Advancing carbohydrate functionality: The role of hypervalent iodine. *Carbohydr. Res.*, 542, 109175, August 2024.
26. Mika, L.T. and Árvai, C., Catalysis in Biomass-Based Solvents, in: *Reference Module in Chemistry, Molecular Sciences and Chemical Engineering*, Elsevier, 2024.
27. Guo, Y., Zhang, J., Wang, C., Liu, M., You, J., Yin, L., *et al.*, Green pretreatment of lignocellulosic biomasses *via* deep eutectic solvents. *Sustain. Chem. Pharm.*, 39, 101569, 2024.
28. Hajba, L. and Guttman, A., Ionic liquids in capillary electrophoresis analysis of proteins and carbohydrates. *J. Chromatogr. A*, 1716, 464642, 8 February 2024.

29. Yang, X., Yan, C., Sun, Y., Liu, Y., Yang, S., Deng, Q., *et al.*, Application of deep eutectic solvents in the pre-processing of atomic spectrometry analysis. *TrAC, Trends Anal. Chem.*, 149, 116555, 2022.
30. Sharma, P. and Zalpouri, R., Microwave-assisted extraction of proteins and carbohydrates from marine resources, in: *Innovative and Emerging Technologies in the Bio-marine Food Sector*, pp. 361–74, 2022.
31. Vatanpour, V., Gul, B.Y., Zeytuncu, B., Korkut, S., İlyasoğlu, G., Turken, T., *et al.*, Polysaccharides in fabrication of membranes: A review. *Carbohydr. Polym.*, 281, 119041, 2022.
32. Lakra, R., Choudhury, S., Basu, S., Recovery of protein and carbohydrate from dairy wastewater using ultrafiltration and forward osmosis processes. *Mater. Today: Proc.*, 47, 1400–3, 2021.
33. Agarwalla, A. and Mohanty, K., A critical review on the application of membrane technology in microalgal harvesting and extraction of value-added products. *Sep. Purif. Technol.*, 344, 127180, 20 September 2024.
34. Smolkova, D., Cmelik, R., Lavicka, J., Labeling strategies for analysis of oligosaccharides and glycans by capillary electrophoresis. *TrAC, Trends Anal. Chem.*, 163, 117068, 2023.
35. Zhao, Y., Ming, Y., Yang, Y., Cai, C., Bi, Y., Fu, Q., *et al.*, Separation of carbohydrates using dynamically adsorbed borate stationary phase for hydrophilic interaction liquid chromatography. *J. Chromatogr. A*, 1720, 464780, 2024.
36. Zhu, L., Zou, S., Yao, D., Li, J., Xiong, Y., Wu, Q., *et al.*, Profiling of aberrant sialylated N- glycans in hepatocellular carcinoma by liquid chromatography mass spectrometry. *Clin. Chim. Acta*, 555, 117827, 2024.
37. Zhang, J.-J., Cheng, F.-F., Zheng, T.-T., Zhu, J.-J., Versatile aptasensor for electrochemical quantification of cell surface glycan and naked-eye tracking glycolytic inhibition in living cells. *Biosens. Bioelectron.*, 89, 937–45, 2017.
38. Capdeville, P., Martin, L., Cholet, S., Damont, A., Audran, M., Ericsson, M., *et al.*, Evaluation of erythropoietin biosimilars Epotin™, Hemax® and Jimaixin™ by electrophoretic methods used for doping control analysis and specific N-glycan analysis revealed structural differences from original epoetin alfa drug Eprex®. *J. Pharm. Biomed. Anal.*, 194, 113750, 2021.
39. Paleček, E. and Černocká, H., Recent advances in electrochemical analysis of biomacromolecules. Catalytic hydrogen evolution and modification of glycoproteins and glycans by Os (VI) complexes. *Curr. Opin. Electrochem.*, 12, 33–41, 2018.
40. Cajic, S., Hennig, R., Grote, V., Reichl, U., Rapp, E., Removable Dyes—The Missing Link for In-Depth N-Glycan Analysis *via* Multi-Method Approaches. *Engineering*, 26, 132–50, 2023.

7

Sustainable Fermentation

Shobika S. and Vijayakumar L.*

Department of Pharmaceutical Technology, Hindusthan Institute of Technology, Coimbatore, India

Abstract

Sustainable microbial fermentation processes offer a promising approach for the production of a wide range of valuable bio-based products, including organic acids, enzymes, fragrances, and pigments. This review provides an overview of the current state of research in this field. Organic acids such as citric, gluconic, lactic, and propionic acid can be effectively produced through fermentative processes, often leveraging renewable and waste-derived feedstocks. Enzymatic production of these organic acids has been a particular focus, with advances in strain development, fermentation optimization, and downstream processing. Beyond organic acids, microbial fermentation has also enabled the sustainable production of other high-value compounds. Solid-state fermentation (SSF) has emerged as an innovative paradigm for valorizing organic waste streams, facilitating the generation of enzymes, biopesticides, biofuels, and even humic acids. The review highlights the key advantages of sustainable microbial fermentation, including the ability to utilize renewable resources, minimize waste, and produce a diverse portfolio of bio-based products. This enhanced overview emphasizes the versatility and sustainability of microbial fermentation processes. By leveraging renewable feedstocks and innovative fermentation techniques such as SSF, the production of organic acids, enzymes, fragrances, and pigments can significantly contribute to a circular bio-based economy.

Keywords: Sustainable fermentation, solid-state fermentation (SSF), organic acids, enzymes, fragrances, pigments

**Corresponding author*: lrvijayakumaar@gmail.com

Senthilkumar Rathnasamy, Vivek Rangarajan and Umile Gianfranco Spizzirri (eds.) Green Engineering and Sustainable Technology: Tools and Techniques, (195–214) © 2026 Scrivener Publishing LLC

7.1 Introduction

Sustainable fermentation technologies have emerged as a promising solution to address the environmental and economic challenges posed by industrialization and the growing global population [1]. These innovative approaches provide a viable pathway for producing a wide range of bio-based products, including biofuels, value-added chemicals, and food ingredients, by utilizing renewable sources instead of fossil-based resources. The significance of sustainable fermentation lies not only in its environmental benefits but also in its potential to drive economic growth in a more sustainable and resilient manner. This chapter offers a comprehensive overview of the principles and applications of sustainable fermentation technology, with a focus on the production of organic acids, enzymes, secondary metabolites, pigments, and fragrance compounds. The aim is to highlight the environmental and economic advantages of these sustainable approaches, as well as to discuss future trends and potential innovations in this rapidly evolving field.

One of the most compelling examples of sustainable fermentation technology is the production of organic acids [21]. Organic acids such as lactic acid (LA), citric acid (CA), and gluconic acid play a crucial role in the food, pharmaceutical, and chemical industries. The microbial production of these acids through fermentation significantly reduces greenhouse gas emissions and often yields products of higher purity compared to traditional chemical synthesis methods. Moreover, advancements in genetic engineering and fermentation optimization have led to substantial improvements in yields and process efficiencies, further enhancing the feasibility and attractiveness of these biotechnological processes [2].

In the realm of enzymes, sustainable fermentation technologies have proven to be indispensable. Enzymes produced through microbial fermentation are essential in a variety of industrial processes, including food processing, biofuel production, and waste management. These biocatalysts offer a sustainable alternative to chemical catalysts, often operating under milder conditions and reducing the need for harsh chemicals. The versatility and efficiency of microbial fermentation for enzyme production underscore its importance in creating more sustainable industrial processes.

Secondary metabolites, such as antibiotics and antifungals, are another significant area where sustainable fermentation technologies excel. Compounds like penicillin and griseofulvin are produced through the fermentation of nutrient-rich media by fungi such as *Penicillium* species [16]. These secondary metabolites are vital in the pharmaceutical industry for treating bacterial and fungal infections. The fermentation-based production of these compounds not only ensures a sustainable supply but also minimizes waste generation and environmental impact, aligning with the principles of green chemistry and sustainable development.

The production of pigments and fragrance compounds through fermentation is yet another testament to the versatility and sustainability of this approach. Pigments produced by fermenting carbohydrate-rich or plant-based feedstocks using microorganisms like *Blakeslea trispora* and *Aspergillus* species find extensive use in the cosmetic, food, and fragrance industries. The sustainable production of these pigments is environmentally friendly and can be further optimized through advancements in microbial strain development and fermentation techniques. These innovations not only enhance the yield and quality of the pigments but also reduce the environmental footprint of the production process.

Sustainable fermentation technologies offer a multifaceted solution to contemporary environmental and economic challenges. They provide a more sustainable alternative to traditional extraction or chemical synthesis methods by utilizing renewable feedstocks [16]. By leveraging innovative microbial processes, these technologies enable the production of a diverse portfolio of bio-based products, significantly contributing to a circular bio-based economy. This chapter seeks to provide an in-depth overview of the principles, applications, and future directions of sustainable fermentation technology, emphasizing its crucial role in promoting a more sustainable and robust industrial framework.

The potential of sustainable fermentation technologies extends beyond their current applications. Ongoing research and development in this field continue to push the boundaries of what is possible, exploring new feedstocks, improving microbial strains, and optimizing fermentation conditions. These advancements promise to unlock even greater efficiencies and open up new avenues for the production of bio-based products. As the global community increasingly recognizes the need for sustainable development, the importance of sustainable fermentation technologies will only grow, positioning them as a cornerstone of the bio-based economy of the future.

Sustainable fermentation technologies represent a powerful tool for addressing some of the most pressing environmental and economic challenges of our time. By harnessing the power of microorganisms and renewable resources, these technologies offer a path toward a more sustainable, efficient, and resilient industrial system. This chapter aims to illuminate the principles and potential of sustainable fermentation, providing a foundation for future innovation and development in this vital field. Figure 7.1 illustrates the diverse products that can be obtained through sustainable fermentation.

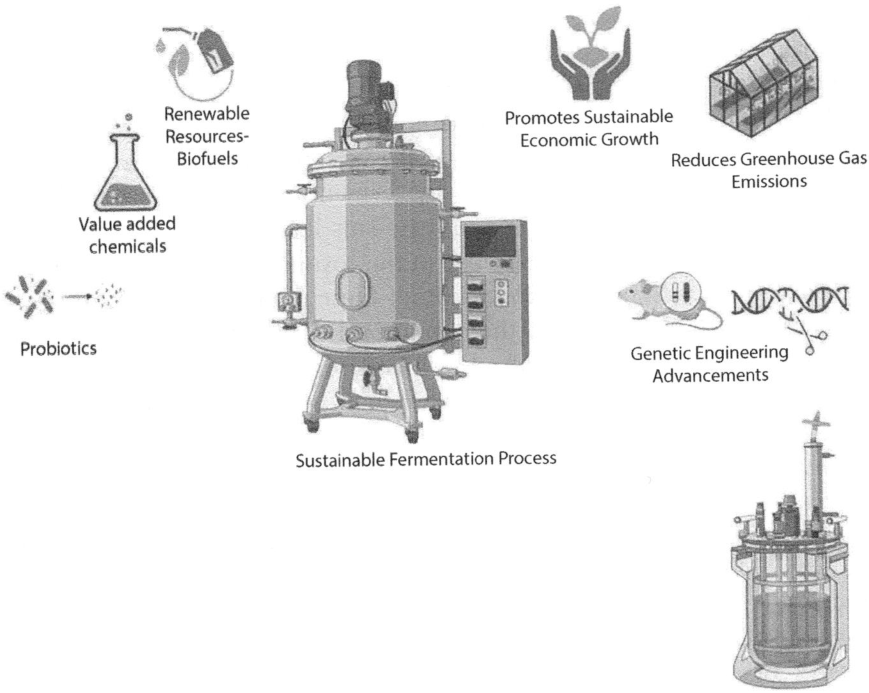

Figure 7.1 Various products derived through sustainable fermentation.

7.2 Organic Acids

7.2.1 Lactic Acid (LA)

LA can be sustainably produced through the fermentation of carbohydrate-rich feedstocks, including glucose, sucrose, and lignocellulosic biomass, using LA bacteria (LAB) such as *Lactobacillus* and *Streptococcus* species. The production process involves several key steps: pre-treatment and hydrolysis of the feedstock to release fermentable sugars, fermentation of these sugars by LAB, and downstream processing for the separation and purification of LA. Pre-treatment methods—mechanical, chemical, or enzymatic—break down complex biomass into simpler sugars. During fermentation, LAB efficiently convert these sugars into LA under controlled conditions. The downstream processing, which includes filtration, precipitation, and distillation, ensures the LA reaches the necessary purity for various applications [5, 22, 38, 41].

LA has diverse applications across the food, pharmaceutical, and chemical industries, serving as a preservative, acidulant, and flavoring agent, and as a precursor for biodegradable polylactic acid polymers [5, 12, 23, 38, 41]. Compared to traditional chemical synthesis, fermentation-based production of LA is more sustainable, utilizing renewable feedstocks such as corn and sugarcane, reducing reliance on fossil fuels, and generating fewer greenhouse gas emissions. Advances in biotechnology, such as the development of robust microbial strains and optimized fermentation conditions, have lowered production costs, improving economic viability. This method also enhances process efficiency through genetic engineering and metabolic pathway optimization and supports a circular economy by utilizing industrial by-products. Furthermore, LA fermentation aligns with regulatory and consumer demands for eco-friendly products. Ongoing research and development in industrial biotechnology continue to improve the sustainability and efficiency of this production method [12, 23, 38].

7.2.2 Citric Acid (CA)

CA is primarily produced through the fermentation of glucose or sucrose using the fungus *Aspergillus niger.* The production process involves several steps: pre-treatment and sterilization of the feedstock, fermentation of the sugars by *A. niger* under controlled conditions, and downstream processing for the extraction and purification of CA [13, 14]. CA is widely used as a flavoring agent, preservative, and pH regulator in the food and beverage industry. It also has applications in the pharmaceutical, cosmetic, and cleaning product industries.

The fermentation-based production of CA is more sustainable compared to traditional chemical synthesis, as it utilizes renewable feedstocks and generates fewer waste products. This method has a lower environmental impact and is more cost-effective due to reduced energy consumption and waste management requirements. Advances in biotechnology, including the optimization of *A. niger* strains and fermentation conditions, have further improved the efficiency and yield of CA production. The integration of these advancements supports a circular economy by maximizing the utilization of feedstocks and minimizing waste, aligning with both regulatory and consumer demands for sustainable and eco-friendly products [1, 14].

7.2.3 Glucaric Acid (GA)

Glucaric acid (GA) can be produced through the fermentation of glucose or other carbohydrate-rich feedstocks using genetically engineered microorganisms, such as *Escherichia coli* or *Saccharomyces cerevisiae* [42].

The production of GA through fermentation involves several key steps. First, microorganisms such as *Escherichia coli* or *Saccharomyces cerevisiae* are genetically engineered to enhance their ability to produce GA by introducing specific genes and optimizing metabolic pathways. During fermentation, these engineered microorganisms metabolize carbohydrate-rich feedstocks like glucose under controlled conditions to maximize GA yield. After fermentation, downstream processing techniques, including filtration, chromatography, and crystallization, are employed to separate and purify the GA. This biotechnological approach is environmentally friendly, as it utilizes renewable feedstocks and reduces greenhouse gas emissions, whereas advancements in genetic engineering and fermentation processes have improved production efficiency and cost-effectiveness [42]. GA has a wide range of potential applications, including the production of biodegradable polymers, detergents, and pharmaceuticals. Additionally, it is being explored as a platform chemical for synthesizing other valuable compounds [39, 42]. It is also being explored as a platform chemical for the synthesis of other valuable compounds.

The fermentation-based production of GA offers significant sustainability advantages over traditional chemical synthesis. By using renewable feedstocks, this method reduces reliance on non-renewable resources and minimizes waste generation. Additionally, the environmental impact is considerably lower, as the fermentation process emits fewer greenhouse gases and pollutants. Continuous advancements in genetic engineering and fermentation technology further enhance the process, improving efficiency and yield, making it a more eco-friendly and economically viable option [39, 42].

7.3 Enzymes

7.3.1 Amylases

Amylases, which catalyze the hydrolysis of starch, are produced through the fermentation of starch-rich feedstocks using microorganisms such as *Bacillus* species or *Aspergillus* species. The production process involves pretreatment and sterilization of the feedstock, followed by fermentation with the amylase-producing microorganisms. After fermentation, downstream processing steps, including filtration, centrifugation, and purification techniques, are employed to extract and purify the amylase enzyme for industrial used [2, 31]. Amylases are utilized across various industries, including food, textile, and paper, for hydrolyzing starch-based materials. In the food industry, they aid in processing and converting starches into simpler sugars. In textiles and paper, amylases enhance the efficiency of starch-based adhesives and coatings. Additionally, they play a crucial role in producing biofuels, such as bioethanol, by breaking down starch-rich feedstocks into fermentable sugars [31]. The fermentation-based production of amylases is a more sustainable approach compared to traditional chemical methods, as it relies on renewable feedstocks and produces fewer waste products. This method not only minimizes environmental impact by reducing emissions and resource depletion but also enhances process efficiency. Further advancements in microbial strain development and fermentation technology are continually improving the yield and sustainability of amylase production. These innovations allow for more precise control of fermentation conditions, optimizing enzyme activity and overall production efficiency. As a result, the fermentation- based approach represents a greener, more efficient alternative to conventional chemical processes [7, 31, 32].

7.3.2 Proteases

Proteases, enzymes that catalyze the hydrolysis of proteins, are produced through the fermentation of protein-rich feedstocks using microorganisms such as *Bacillus* species or *Aspergillus* species. These microorganisms are cultured under controlled conditions to optimize enzyme production, facilitating the breakdown of proteins into peptides and amino acids [27]. The production of proteases through fermentation involves several key steps. First, the feedstock, which is rich in proteins, undergoes pretreatment and sterilization to prepare it for microbial fermentation. This initial step ensures that the feedstock is free from contaminants and optimally conditioned for enzyme production. Next, the prepared feedstock is fermented using protease-producing microorganisms such as *Bacillus* species or *Aspergillus* species. During fermentation, these microorganisms break down the protein substrates into peptides and amino acids, producing protease enzymes. Following fermentation, the downstream processing phase begins, which includes the extraction and purification of the protease enzyme. This phase typically involves filtration to remove microbial cells and debris, followed by various purification techniques such as ion exchange chromatography, precipitation, and concentration to isolate and purify the protease enzyme to the desired quality. These steps ensure that the protease is produced efficiently and is suitable for various industrial applications [27]. Proteases have a wide range of applications in the food, detergent, and pharmaceutical industries, where they are used for the hydrolysis of proteins. They are also used in the production of protein hydrolysates and peptides for various applications [33].

The fermentation-based production of proteases is more sustainable than traditional chemical methods, as it leverages renewable feedstocks and produces fewer waste products. This approach not only reduces the overall environmental impact but also enhances efficiency. Additionally, ongoing advancements in microbial strain development and fermentation technology continue to optimize the process, improving enzyme yield and reducing resource consumption. This makes the fermentation-based method a greener and more effective alternative to conventional chemical synthesis [11, 27].

7.4 Secondary Metabolites

7.4.1 Penicillin

Penicillin, a widely utilized antibiotic, is produced through the fermentation of the fungus *Penicillium chrysogenum*. This process involves cultivating the fungus in a controlled environment where it produces penicillin as a metabolic by-product [16]. The production of penicillin involves several key steps. First, the feedstock, which often includes nutrient-rich media such as corn steep liquor, undergoes pretreatment and sterilization to ensure it is free from contaminants and suitable for microbial growth. Next, *Penicillium chrysogenum* is used to ferment this sterilized feedstock under carefully controlled conditions, allowing the fungus to produce penicillin as a by-product. After fermentation, the process moves to downstream processing, where the penicillin is extracted from the fermentation broth and purified through various techniques to obtain the final antibiotic product [15]. It is widely used in the pharmaceutical industry for the treatment of bacterial infections. It is one of the most important and widely used antibiotics in modern medicine. The fermentation-based production of penicillin presents a more sustainable option compared to traditional chemical synthesis. This method utilizes renewable feedstocks, such as corn steep liquor and other nutrient-rich media, which reduces dependence on non-renewable resources and results in less waste generation.

Additionally, the fermentation process has a lower environmental impact by emitting fewer pollutants and greenhouse gases. Advancements in microbial strain development and fermentation technology further enhance the efficiency of penicillin production, leading to higher yields and reduced energy consumption. These improvements not only contribute to a more sustainable production process but also offer economic benefits by lowering production costs and resource usage [6, 8].

7.4.2 Griseofulvin

Griseofulvin, an antifungal medication, is produced by fermenting the fungus *Penicillium griseofulvum*. This method leverages the natural metabolic processes of the fungus to efficiently generate the compound used in treating fungal infections [3, 18]. The production of griseofulvin involves several key steps. Initially, the feedstock, which may include corn steep liquor or other nutrient-rich media, undergoes pre-treatment and sterilization to ensure it is suitable for microbial growth. Next, *Penicillium griseofulvum* is used to ferment the prepared feedstock under controlled conditions, allowing the fungus to produce griseofulvin. After fermentation, the process includes downstream processing steps such as extraction and purification to isolate the griseofulvin. This antifungal medication is widely utilized in the pharmaceutical industry to treat fungal infections, particularly those affecting the skin, hair, and nails [3, 18].

The fermentation-based production of griseofulvin is more sustainable than traditional chemical synthesis because it utilizes renewable feedstocks and produces fewer waste products. This method not only has a lower environmental impact but also benefits from ongoing advancements in microbial strain development and fermentation technology, which further enhance process efficiency and reduce resource consumption. These innovations contribute to a more eco-friendly and cost-effective production approach [3].

7.5 Pigments

7.5.1 Carotenoid

Carotenoids, a group of naturally occurring pigments, are produced through the fermentation of various microorganisms such as *Blakeslea trispora*, *Dunaliella salina*, or *Phaffia rhodozyma* [25, 29, 40]. These microorganisms are cultured under specific conditions that promote the synthesis of carotenoids, which are then harvested and purified for various applications. The production process for carotenoids involves several critical steps. Initially, the feedstock—such as glucose, glycerol, or other carbon sources—is pretreated and sterilized to ensure it is free of contaminants and suitable for microbial fermentation. This step prepares the feedstock for optimal microbial activity. Subsequently, the sterilized feedstock is used in the fermentation process with microorganisms like *Blakeslea trispora*, *Dunaliella salina*, or *Phaffia rhodozyma*, yeast species such as Rhodotorula have also been identified as efficient microbial producers of carotenoids, with recent advances reported in fermentation-based production [24, 28], which are cultured under controlled conditions to produce carotenoids. Finally, after fermentation, downstream processing is carried out to extract and purify the carotenoids, ensuring they meet the required standards for various applications [25, 29]. Carotenoids have a wide range of applications across the food, cosmetic, and pharmaceutical industries. They are utilized as natural colorants, providing vibrant hues to various products. Additionally, carotenoids serve as antioxidants, helping to protect cells from damage caused by free radicals. They are also employed as nutritional supplements, offering health benefits such as improved vision and immune support [25].

The fermentation-based production of carotenoids is more sustainable than traditional chemical synthesis or extraction from natural sources, as it relies on renewable feedstocks and generates fewer waste products. This method also results in a lower environmental impact by reducing the consumption of non-renewable resources and minimizing pollutants. Additionally, advancements in microbial strain development and fermentation technology continue to enhance process efficiency and yield, making the production of carotenoids more eco-friendly and economically viable [34].

7.5.2 Anthocyanins

Anthocyanins, a type of water-soluble pigment, can be produced through the fermentation of plant-based feedstocks, including fruits and vegetables, using microorganisms such as *Aspergillus* or *Penicillium* species [26]. The production of anthocyanins begins with the pretreatment and sterilization of plant-based feedstocks, such as fruits and vegetables, to prepare them for fermentation. This step is crucial to eliminate contaminants and optimize conditions for microbial activity. Following pretreatment, the feedstock undergoes fermentation using anthocyanin-producing microorganisms, such as *Aspergillus* or *Penicillium* species, under carefully controlled conditions to maximize pigment production. After fermentation, downstream processing involves the extraction and purification of anthocyanins to achieve the desired concentration and purity. Techniques such as filtration, centrifugation, and chromatography are often employed to isolate and refine the pigments. These high-purity anthocyanins are then used in various applications across the food, cosmetic, and pharmaceutical industries due to their natural colorant properties, antioxidant benefits, and potential health advantages [5].

The fermentation-based production of anthocyanins is more sustainable than traditional extraction from natural sources, as it utilizes renewable feedstocks and generates fewer waste products. This method reduces reliance on natural plant sources, which can be limited and environmentally taxing to harvest. Additionally, the process has a lower environmental impact and can be enhanced through advancements in microbial strain development and fermentation technology, leading to improved yields and efficiency. These innovations contribute to a more eco-friendly and cost-effective production method [19].

7.6 Fragrance Compounds

7.6.1 Vanillin

Vanillin, the main component of natural vanilla flavor, can be produced through the fermentation of feedstocks such as ferulic acid or eugenol using genetically engineered microorganisms like *Escherichia coli* or *Saccharomyces cerevisiae*. The primary production process includes genetic engineering of these microorganisms to enhance vanillin production, ensuring higher yields and efficiency [5, 10]. The production of vanillin involves several key steps. Initially, genetic engineering is used to modify microorganisms such as *Escherichia coli* or *Saccharomyces cerevisiae* to enhance their vanillin production capabilities. This involves inserting specific genes responsible for converting feedstocks like ferulic acid or eugenol into vanillin. The engineered microorganisms are then used to ferment the feedstock under controlled conditions to optimize vanillin yield. Following fermentation, downstream processing steps, including extraction and purification, are carried out to isolate and refine the vanillin for use. This method ensures a sustainable, high-yield production process. Vanillin is widely used as a flavoring agent in the food, beverage, and cosmetic industries [35]. It is also used in the production of various pharmaceutical and chemical compounds.

The fermentation-based production of vanillin is more sustainable compared to traditional extraction from natural sources or chemical synthesis, as it utilizes renewable feedstocks and generates fewer waste products. This method reduces environmental impact and can be further enhanced through advancements in genetic engineering and fermentation technology, leading to higher efficiency and yield. These improvements contribute to a more eco-friendly and cost- effective production process [10, 20].

7.6.2 Limonene

Limonene, a naturally occurring terpene with a citrus-like aroma, can be produced through the fermentation of various carbohydrate-rich feedstocks using genetically engineered microorganisms, such as *Escherichia coli* or *Saccharomyces cerevisiae.* The production process of limonene begins with the genetic engineering of microorganisms, such as *Escherichia coli* or *Saccharomyces cerevisiae*, to enhance their capability to produce limonene. This involves incorporating specific genes responsible for the biosynthesis of limonene, optimizing the metabolic pathways within the microorganism. Once genetically engineered, these microorganisms are used to ferment the feedstock under controlled conditions, ensuring optimal growth and limonene production. During fermentation, the microorganisms convert the feedstock into limonene through their metabolic activities. Following fermentation, downstream processing steps are implemented, which include the extraction and purification of limonene to achieve the desired purity and concentration. This biotechnological approach not only ensures a sustainable and efficient production process but also allows for scalable and environmentally friendly manufacturing of limonene. It has a wide range of applications in the food, cosmetic, and fragrance industries, where it is used as a flavoring agent and fragrance compound [17]. It is also used in the production of various chemical intermediates and solvents.

The fermentation-based production of limonene offers a sustainable alternative to traditional extraction from natural sources or chemical synthesis, as it employs renewable feedstocks and produces fewer waste products. This approach alleviates the strain on citrus resources, reducing environmental impact. Furthermore, the process benefits from lower ecological footprints and can be significantly enhanced through advancements in genetic engineering and fermentation technologies. These improvements boost production efficiency and yield, making the method both eco-friendly and economically attractive [36].

7.7 Comparative Analysis and Sustainability

Sustainable fermentation technologies offer significant environmental and economic benefits compared to traditional production methods. Table 7.1 provides an overview of various byproducts, their microbial sources,

Table 7.1 Overview of various bioproducts and their production processes.

Bioproduct	Production process	Organism	Yield (%)	References
Organic acids				
Lactic acid (LA)	Submerged fermentation of carbohydrate-rich feedstocks (e.g., glucose, sucrose, and lignocellulosic biomass) by lactic acid bacteria (LAB)	*Lactobacillus* and *Streptococcus*	33% 90%–95%	[37]
Citric acid (CA)	Submerged fermentation of glucose or sucrose by the fungus *Aspergillus niger*	*Aspergillus niger*	60%–80%	[13]
Glucaric acid (GA)	Fermentation of glucose or other carbohydrates using genetically engineered microorganisms	*E. coli* and *S. cerevisiae*	50%–70%	[39, 42]
Enzymes				
Amylases	Submerged fermentation of starch-rich feedstocks	*Bacillus* and *Aspergillus*	60%–80%	[2, 7, 31]
Proteases	Submerged fermentation of protein-rich feedstocks (including agricultural waste)	*Bacillus*, *Aspergillus*, and other engineered strains	Can exceed 70%	[11, 31]

(*Continued*)

Table 7.1 Overview of various bioproducts and their production processes. (*Continued*)

Bioproduct	Production process	Organism	Yield (%)	References
Secondary metabolites				
Penicillin	Submerged fermentation of nutrient-rich media	*Penicillium chrysogenum*	Can exceed 80%	[6]
Griseofulvin	Submerged fermentation of nutrient-rich media	*Penicillium griseofulvum*	40%–60%	[3]
Pigments				
Carotenoids	Submerged fermentation of carbohydrate-rich feedstocks	*B. trispora, D. salina*, and *P. rhodozyma*	30%–50%	[22, 24]
Anthocyanins	Submerged fermentation of plant-based feedstocks	*Aspergillus* and *Penicillium*	20%–40%	[4, 6]
Fragrance compounds				
Vanillin	Fermentation of ferulic acid or eugenol using genetically engineered microorganisms	*E. coli* and *S. cerevisiae*	40%–60%	[9]
Limonene	Fermentation of carbohydrate-rich feedstocks using genetically engineered microorganisms	*E. coli* and *S. cerevisiae*	30%–50%	[17]

production processes, and yields. These technologies utilize a diverse range of renewable feedstocks, such as agricultural residues, food waste, and lignocellulosic biomass, reducing reliance on fossil-based resources and promoting resource efficiency [30, 37]. Fermentation processes generate fewer waste products and by-products, leading to a more efficient and environmentally friendly production cycle [9]. Additionally, they have a lower carbon impact than conventional chemical synthesis, contributing to the reduction of greenhouse gas emissions and aiding in climate change mitigation. Economic viability is enhanced through continuous advancements in fermentation technology, microbial strain development, and the utilization of low-cost feedstocks, making these sustainable methods more cost-effective and enabling their widespread adoption. These innovations collectively highlight the transformative potential of fermentation-based technologies in fostering a sustainable and economically resilient industrial landscape [22].

7.8 Conclusion

This chapter has provided a comprehensive overview of the principles and applications of sustainable fermentation technology, encompassing a wide range of products, including organic acids, enzymes, secondary metabolites, pigments, and fragrance compounds. The comparative analysis has demonstrated the significant environmental and economic benefits of these sustainable approaches over traditional production methods. The successful implementation of sustainable fermentation technology is crucial in addressing global challenges such as resource depletion, environmental degradation, and the urgent need for sustainable production processes. By utilizing microorganisms and renewable feedstocks, fermentation-based production methods contribute to the development of a circular and sustainable economy, reducing reliance on non-renewable resources and minimizing waste generation and greenhouse gas emissions.

Future trends in sustainable fermentation technology include the development of more efficient microbial strains through genetic engineering, optimization of fermentation processes, and integration of advanced separation and purification techniques. These advancements will enhance the sustainability and economic viability of fermentation-based production methods, expanding their application to a broader range of bioproducts. Interdisciplinary collaboration and integration of systems biology, synthetic biology, and metabolic engineering will be essential in overcoming current limitations and accelerating the transition to a bio-based economy.

By pushing the boundaries of innovation and sustainability, fermentation-based production has the potential to revolutionize industrial processes and contribute significantly to global sustainability goals.

AI disclosure statement: The author acknowledges the use of AI-based tools for minor language editing or grammatical correction. However, all intellectual contributions, analyses, and conclusions presented in this chapter are entirely the author's own.

References

1. Aynalem, B., Negi, H., Alemu, Y., Sehrawat, N., Kumar, A., Citric acid: fermentative production using organic wastes as feedstocks. *Phys. Sci. Rev.*, 9, 9, 3043–3055, 2023.
2. Ayinde, E.O., Adejumo, T.O., Adeoyo, O.R., Production, optimization and characterization of amylase using solid state fermentation method. *Front. Environ. Microbiol.*, 9, 1, 9, 2023.
3. Aris, P., Wei, Y., Mohamadzadeh, M., Xia, X., Griseofulvin: An updated overview of old and current knowledge. *Molecules/Molecules Online/Molecules Annual*, *27*, 20, 7034, 2022.
4. Cappellini, F., Marinelli, A., Toccaceli, M., Tonelli, C., Petroni, K., Anthocyanins: From mechanisms of regulation in plants to health benefits in foods. *Front. Plant Sci.*, *12*, 2021, https://doi.org/10.3389/fpls.2021.748049.
5. Daful, A.G., Loridon, M., Chandraratne, M.R., Lactic Acid Production from Lignocellulosic Biomass, in: 10.5772/intechopen.112739, 2023.
6. Fatima, S., Rasool, A., Sajjad, N., Bhat, E.A., Hanafiah, M.M., Mahboob, M., Analysis and evaluation of penicillin production by using soil fungi. *Biocatal. Agric. Biotechnol.*, *21*, 101330, 2019.
7. Fazil, N.M.M., Javed, N.I., Ali, N.K., Waheed, N.H., Dastagir, N.N., Production optimization and industrial applications of amylase from indigenous bacterial species using banana peels. *Biosight J.*, *4*, 1, 41–51, 2023, https://doi.org/10.46568/bios.v4i1.75.
8. Ismalaj, E. and De Borggraeve, W., Penicillins, in: pp. 116–158, Elsevier eBooks, 2022, 10.1016/B978-0-12-818655-8.00141-4.
9. Jain, S., Heffernan, J., Joshi, J., Watts, T., Marcellin, E., Greening, C., Microbial conversion of waste gases into single-cell protein. *Microbiol. Aust.*, *44*, 1, 27–30, 2023, https://doi.org/10.1071/ma23007.
10. Jiang, W., Chen, X., Feng, Y., Sun, J., Jiang, Y., Zhang, W., Xin, F., Jiang, M., Current status, challenges, and prospects for the biological production of vanillin. *Fermentation*, *9*, 4, 389, 2023.
11. Jamal, G.A., Jahangirian, E., Hamblin, M.R., Mirzaei, H., Tarrahimofrad, H., Alikowsarzadeh, N., Proteases, a powerful biochemical tool in the service

of medicine, clinical and pharmaceutical. *Prep. Biochem. Biotechnol.*, 55, 1, 1–25, 2024.
12. Kim, J., Kim, Y., Lebaka, V.R., Wee, Y., Lactic Acid for green chemical industry: Recent advances in and future prospects for production technology, recovery, and applications. *Fermentation*, 8, 11, 609, 2022.
13. Khurshid, S., Ashraf, H., Hussain, T., Iqbal, M., Qureshi, H., Anwar, T., Salmen, S.H., Ansari, M.J., Enhanced Citric Acid Production through Aspergillus niger: Insights from Fermentation Studies Using Sugarcane Molasses. *Life*, *14*, 6, 756, 2024, https://doi.org/10.3390/life14060756.
14. Kumar, A., Aynalem, B., Alemu, Y., Negi, H., Sehrawat, N., 2 Citric acid: fermentative production using organic wastes as feedstocks, in: pp. 21–34, De Gruyter eBooks, 2024. 10.1515/9783110792584-002.
15. Khousin, S., Dutta, S., Kiruthika, S.A., Comparative analysis of penicillin production using various media compositions: A research study. *Int. J. Res. Publ. Rev.*, *4*, 9, 1936–1941, 2023.
16. Kosalková, K., Sánchez-Orejas, I., Cueto, L., García-Estrada, C., Penicillium chrysogenum Fermentation and Analysis of Benzylpenicillin by Bioassay and HPLC, in: *Methods in molecular biology*, pp. 195–207, 2021.
17. Kong, X., Wu, Y., Yu, W., Liu, Y., Li, J., Du, G., Lv, X., Liu, L., Efficient Synthesis of Limonene in Saccharomyces cerevisiae Using Combinatorial Metabolic Engineering Strategies. *J. Agric. Food. Chem.*, *71*, 20, 7752–7764, 2023.
18. Lv, J., Yao, L., Li, S., Ye, M., Li, D., Li, C., Li, Y., Three new griseofulvin derivatives from *Aureobasidium pullulans*. *Nat. Prod. Res.*, 1–6, 39, 13, 3745–3750, 2024.
19. Lin, Y., Li, C., Shi, L., Wang, L., Anthocyanins: Modified new technologies and challenges. *Foods*, *12*, 7, 1368, 2023, https://doi.org/10.3390/foods12071368.
20. Liu, Y., Sun, L., Huo, Y., Guo, S., Strategies for improving the production of bio-based vanillin. *Microb. Cell Fact.*, *22*, 1, 2023.
21. Linder, T., Fulfilling the promises of Fermentation-Derived foods. *GEN Biotechnol.*, *2*, 3, 188–196, 2023.
22. Nehra, S., Gothwal, R.K., Varshney, A.K., Meena, P., Trivedi, P.C., Ghosh, P., Carotenoids in microorganisms and their applications, in: pp. 228–239, Bentham Science Publishers eBooks, 2024. 10.2174/9789815079753124030017.
23. Ojo, A.O. and De Smidt, O., Lactic Acid: A Comprehensive review of production to purification. *Processes*, *11*, 3, 688, 2023.
24. Ochoa-Viñals, N., Alonso-Estrada, D., Pacios-Michelena, S., García-Cruz, A., Ramos- González, R., Faife-Pérez, E., Michelena-Álvarez, L.G., Martínez-Hernández, J.L., Iliná, A., Current Advances in Carotenoid Production by Rhodotorula sp. *Fermentation*, *10*, 4, 190, 2024.
25. Pappalardo, I., Convertini, P., Infantino, V., Anthocyanins, in: pp. 1221–1239, Elsevier eBooks, 2024, https://doi.org/10.1016/b978-0-443-23763-8.00024-5.
26. Ratnaningrum, D., Kosasih, W., Endah, E.S., Lathifa, A.K.N., Diwan, A.M., Nida, V., Saraswaty, V., Risdian, C., Protease production by soil bacteria for

green technology: Screening and optimization. *IOP Conf. Ser. Earth Environ. Sci., 1201*, 1, 012094, 2023.

27. Rifaie, R.M.M. and Karim, L., Production of Vanillin from Pumpkin Peels *via* Microbiological Fermentation using Aspergillus niger. *Malays. J. Fundam. Appl. Sci., 19*, 6, 1119–1130, 2023b.
28. Rodríguez, N.A.T., Quiñones-Cerna, C.E., Castillo, H.M.R., Cruz-Monzon, J.A., Butrón, F.J.H., Soto, J.C.R., Optimization of Total Carotenoid Production by Rhodotorula mucilaginosa from Artichoke Agroindustrial Waste Using Response Surface Methodology. *Environ. Res. Eng. Manage., 79*, 2, 111–121, 2023.
29. Ruslan, N.F., *et al.*, Sustainable bioethanol production by solid-state fermentation: a systematic review. *Environ. Sci. Pollut. Res. Int.*, 32, 22, 13140–13158, 2025. doi:10.1007/s11356-024-35406-z.
30. Srivastav, Y., Thakur, G., Srivastav, A., Starch Hydrolyzing Enzyme Production from Fruit Peels using Aspergillus Niger. *Biotechnol. J. Int., 28*, 4, 50–70, 2024.
31. Shad, M., Hussain, N., Usman, M., Akhtar, M.W., Sajjad, M., Exploration of computational approaches to predict the structural features and recent trends in α- amylase production for industrial applications. *Biotechnol. Bioeng., 120*, 8, 2092–2116, 2023.
32. Song, P., Zhang, X., Wang, S., Xu, W., Wang, F., Fu, R., Wei, F., Microbial proteases and their applications. *Front. Microbiol., 14*, 2023b, https://doi.org/10.3389/fmicb.2023.1236368.
33. Su, B., Deng, M., Zhu, H., Advances in the discovery and engineering of gene targets for carotenoid biosynthesis in recombinant strains. *Biomolecules, 13*, 12, 1747, 2023.
34. Khanijou, J.K., *et al.*, Systems biology approach for enhancing limonene yield by re-engineering Escherichia coli. *npj Syst. Biol. Appl.*, 10, 1, 109, 1 Oct. 2024, doi:10.1038/s41540-024-00440-7.
35. Schempp, F.M., Strobel, I., Etschmann, M.M.W., Bierwirth, E., Panten, J., Schewe, H., Schrader, J., Buchhaupt, M., Identification of fungal limonene-3-hydroxylases for biotechnological menthol production. *bioRxiv (Cold Spring Harbor Laboratory)*, 87, 10, e02873–20, 2020.
36. Singh, K.A., Saroch, A., Rajput, R., Microbial diversity of fermented foods. *Curr. J. Appl. Sci. Technol., 42*, 47, 20–31, 2023, https://doi.org/10.9734/cjast/2023/v42i474313.
37. Yankov, D., Fermentative lactic acid production from lignocellulosic feedstocks: from source to purified product. *Front. Chem.*, 10, 823005, 2022.
38. Yi, Y. and Ng, I., Toward Low-Carbon-Footprint Glycolic Acid Production for Bioplastics through Metabolic Engineering in Escherichia coli. *ACS Sustain. Chem. Eng., 11*, 2, 815–823, 2023.
39. Yang, X., Niu, L., Ye, C., Wang, Y., Liu, Y., Wang, F., Sun, N., Biosynthesis of Glucaric Acid by Recombinant Strain of Escherichia coli Expressing Two Different Urinate Dehydrogenases. *Fermentation, 9*, 8, 764, 2023.

40. Yaderets, V.V., Karpova, N.V., Glagoleva, E.V., Petrova, K.S., Shibaeva, A.S., Dzhavakhiya, V.V., Carotenoids: Overview of the main methods and conditions of their preparation. *Proc. Univ. Appl. Chem. Biotechnol.*, *14*, 1, 41–54, 2024b.
41. Zhao, Z., Wang, C., Di, W., Hao, Y., Shi, S., Zhai, Z., Fermentative production of L-lactic acid from starch-based waste by engineered *Lactiplantibacillus plantarum* CAUH2-4, 2024. 10.21203/rs.3.rs-3802235/v1.
42. Zhao, Y., Zuo, F., Shu, Q., Yang, X., Deng, Y., Efficient Production of Glucaric Acid by Engineered Saccharomyces cerevisiae. *Appl. Environ. Microbiol.*, *89*, 6, 2023.

8

Green Electrolytes

Lavanya Priyadarshini Ramalingam[1], Senthilkumar Rathnasamy[2], Balasubramanian Ramalingam[1] and Parkavi Kathirvelu[1*]

[1]*Electrical and Electronics Engineering, SASTRA Deemed to be University, Thanjavur, India*
[2]*Green Separation Engineering Laboratory, School of Chemical and Biotechnology, SASTRA Deemed to be University, Thanjavur, India*

Abstract
Energy storage has always been an essential factor in contributing to renewable energy avenues. Advancement in this field has revolutionized the adaptation of sustainability and widened the scope of its application. Over the recent decade, sustainability has expanded its presence in energy storage by the introduction of green electrolytes. Beyond non-toxic and biodegradable in nature, these alternative electrolytes have positively influenced the storage devices by increasing their lifetime and expanding their capacity. This chapter unveils the role, mechanism, and consequence of using green electrolytes in conventional batteries and their socio-economic significance. Further, the key aspects of selecting these electrolyte systems are also described in detail. Additionally, recent developments on sustainable gel electrolyte systems have also been outlined. Finally, the enhanced performance of these energy storage devices along with their challenges and prospects has also been listed in this chapter.

Keywords: Electrolyte, deep eutectic solvent, electrochemical energy storage, machine learning, electrochemical performance, eutectogel, life cycle, electric vehicle

8.1 Introduction

The accelerating impacts of climate change, largely driven by the increase in greenhouse gas emissions from human activities, have heightened the

**Corresponding author*: to_parkavi@eee.sastra.edu

Senthilkumar Rathnasamy, Vivek Rangarajan and Umile Gianfranco Spizzirri (eds.) Green Engineering and Sustainable Technology: Tools and Techniques, (215–248) © 2026 Scrivener Publishing LLC

urgency to adopt sustainable energy solutions. Among the primary contributors to environmental degradation are internal combustion engine (ICE) vehicles, which release substantial quantities of carbon dioxide (CO_2) and other pollutants such as nitrogen oxides (NO_x) and particulate matter. These emissions not only exacerbate global warming but also significantly impair air quality, causing widespread health and ecological damage [1]. In response to these pressing issues, the transition to zero-emission vehicles (ZEVs) has gained traction as a pivotal strategy to combat pollution and mitigate climate change. Electric vehicles (EVs) have emerged as the leading candidates for this transition, offering an effective alternative to fossil fuel–powered vehicles by eliminating tailpipe emissions. This shift is closely aligned with global sustainability goals, including achieving carbon neutrality and advancing the United Nations' Sustainable Development Goals (SDGs). However, the transition to electric mobility comes with significant challenges, particularly in the realm of battery technology, which forms the cornerstone of EV performance and sustainability [2].

At the heart of EV batteries lie two critical components: electrodes and electrolytes. Electrolytes are the medium through which ions move between the cathode and anode during charging and discharging cycles, making them indispensable for the battery's operation. The properties of the electrolyte—such as ionic conductivity, stability, and compatibility with electrode materials—directly impact the battery's energy density, safety, cycle life, and overall efficiency [3]. Traditional electrolytes, typically composed of organic or aqueous solvents and dissolved salts, pose several challenges. These solvents are often volatile, flammable, and derived from non-renewable resources, raising concerns about safety and environmental sustainability. Additionally, their disposal and recycling remain problematic due to their toxicity and chemical instability. To address these challenges, researchers have turned their focus toward the development of green electrolytes. Green electrolytes are designed to be eco-friendly, sustainable, and safer, and derived from renewable resources or synthesized using less harmful substances. They aim to reduce the ecological footprint of EV batteries while maintaining or enhancing their performance. Promising candidates for green electrolytes include ionic liquids (ILs), deep eutectic solvents (DESs), solid-state electrolytes (SSEs), and water-based systems [4].

DESs, a mixture of hydrogen bond donors (HBDs) and hydrogen bond acceptors (HBAs), exhibiting low volatility, high thermal stability, and tunable properties, make them particularly attractive for their ease of synthesis, biodegradability, and potential to improve the safety and stability of batteries. ILs, known for their non-volatility and high ionic conductivity, offer improved safety profiles and reduced risks of thermal runaway, although their cost and environmental impact remain areas of ongoing research [5]. SSEs replace liquid electrolytes with solid materials, addressing leakage and flammability issues while enabling higher energy densities and the use of metallic lithium anodes, which can significantly improve battery performance. Aqueous electrolytes, being water-based, are inherently safer and less harmful, and recent advancements in stabilizing these systems have made them viable candidates for certain battery chemistries [6].

Beyond safety and sustainability, green electrolytes contribute to solving key challenges in EV adoption, such as range anxiety—the concern over limited driving range—and prolonged charging times. By improving the ionic conductivity and thermal stability of batteries, green electrolytes enhance energy density and enable faster charge-discharge cycles. These improvements directly translate to longer ranges and shorter charging durations, addressing critical consumer concerns and bolstering EV adoption. Figure 8.1 illustrates the DESs as green electrolytes in lithium-ion batteries (LIBs) for EVs. The development of green electrolytes is also integral to advancing a circular economy. This approach emphasizes resource efficiency, waste reduction, and the reuse of materials throughout their life cycle. By utilizing renewable feedstocks and designing recyclable and biodegradable electrolytes, green chemistry principles are embedded in battery production, minimizing the environmental impact from manufacturing to disposal [7].

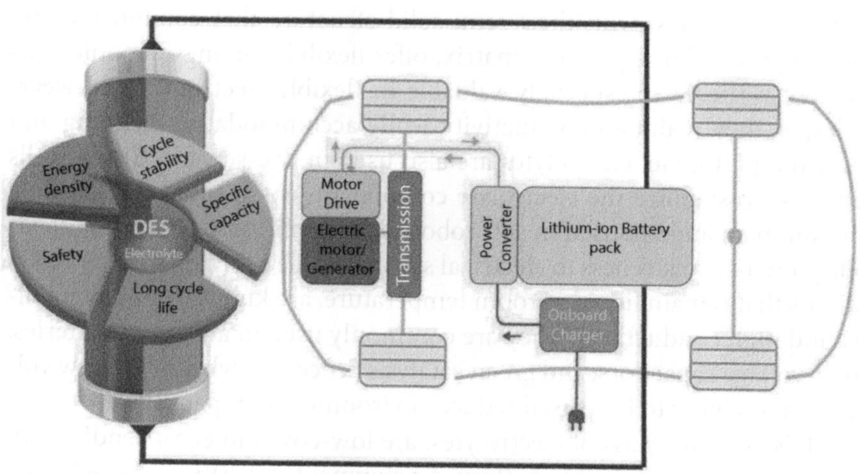

Figure 8.1 Deep eutectic solvents as green electrolytes in lithium-ion batteries for electric vehicles.

Green electrolytes represent a transformative approach to overcoming the limitations of traditional battery chemistries. This chapter delves into their pivotal role in shaping the future of EV technology, providing an overview of recent innovations, the challenges involved, and the prospects for creating batteries that are not only high-performing but also aligned with the imperatives of sustainability and environmental stewardship. By optimizing the composition and functionality of these electrolytes, researchers aim to realize a future where EVs are both an effective and responsible solution to the challenges of energy and mobility in a changing climate [8].

8.1.1 Electrolyte and Its Types

Electrolytes are crucial for conducting electricity by dissociating into ions, making them essential in many fields, especially electrochemical applications. Liquid electrolytes, commonly found in batteries and fuel cells, include aqueous or non-aqueous solutions with dissolved salts like potassium hydroxide (KOH) for alkaline batteries or lithium salts in LIBs. These liquid electrolytes enable ion flow between electrodes, crucial for energy storage and discharge, and are also used in supercapacitors, which require quick charge-discharge cycles for renewable energy storage and EVs. In contrast, solid electrolytes like lithium garnet or polymer-based types offer enhanced safety by reducing risks associated with leakage and flammability. Their stable, solid-state structure makes them ideal for solid-state batteries, sensors, and even medical devices, where stability and longevity are key [9].

Gel electrolytes, with their semi-solid structure that combines liquid electrolytes within a polymer matrix, offer flexibility alongside ionic conductivity. These are especially valuable in flexible electronics and wearables, as they maintain conductivity while accommodating bending and stretching [10]. Gel electrolytes are also used in dye-sensitized solar cells (DSSCs) to stabilize the electrolyte component, enhancing cell efficiency and lifespan, and are used in soft robotics for applications requiring flexibility and responsiveness to electrical stimuli. IL electrolytes, a special class of salts that remain liquid at room temperature, are known for their stability and high conductivity. They are commonly used in advanced batteries, high-voltage capacitors, and green catalysis processes, where their low volatility and ability to be reused reduce environmental impact [11].

DESs, a newer class of electrolytes, are low-cost and eco-friendly, with applications in sustainable battery development for EVs, electroplating, and even biomedical fields, where their biocompatibility is highly valuable. Lastly, molten salt electrolytes, which consist of metal salts that conduct ions at high temperatures, are key to applications requiring thermal stability, such as thermal energy storage in concentrated solar power plants and high-temperature batteries like sodium-sulfur cells, which are useful for grid storage [12]. They are also used in metal processing due to their ability to handle extreme temperatures required for electrochemical metallurgy. Each type of electrolyte serves unique roles based on its properties, from sustainable energy to environmental and biomedical applications, with ongoing research focused on enhancing performance, safety, and environmental sustainability [13].

8.1.2 The Evolution of Green Electrolytes

The advancement of green electrolytes, as depicted in Figure 8.2, highlights their pivotal role in enhancing energy storage performance while promoting environmental sustainability for EVs. Initially, the electrolytes used in early batteries such as lead-acid and nickel-cadmium batteries were primarily aqueous solutions. These water-based electrolytes were functional and effective in their time, but they posed significant limitations, including a narrow electrochemical window, instability at higher voltages, and corrosiveness. As technology advanced, particularly with the rise of consumer electronics and EVs, these traditional electrolytes could no longer meet the demands for higher energy densities and longer life cycles.

With the advent of LIBs in the late 20th century, organic liquid electrolytes became the standard. These electrolytes, typically containing lithium salts like lithium hexafluorophosphate ($LiPF_6$) in organic carbonate

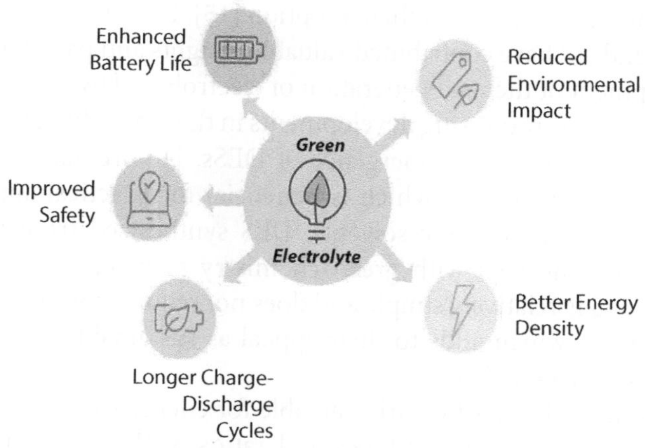

Figure 8.2 Advantages of green electrolytes for electric vehicles.

solvents, represented a substantial leap in energy density and overall battery performance. They facilitated the mass adoption of portable electronics and laid the groundwork for the EV industry. However, despite their success, these organic electrolytes introduced a new set of challenges. They are flammable, toxic, and pose significant environmental hazards when disposed of, as they are neither biodegradable nor easy to recycle. Their high volatility and potential to cause thermal runaway—a rapid increase in temperature that can lead to fires or explosions—became major safety concerns, particularly as batteries scaled up in size and energy content.

In response to these challenges, research began to pivot toward safer and more sustainable alternatives, leading to the exploration of SSEs and ILs. Solid-state electrolytes, as the name suggests, are materials that conduct ions in a solid form rather than a liquid. They offer significant safety advantages, including lower risks of leakage, flammability, and volatility, making them particularly attractive for EVs and other high-energy applications [14]. However, SSEs also have limitations, such as lower ionic conductivity compared to liquid electrolytes, especially at room temperature. This conductivity issue has been a bottleneck for large-scale adoption, as it hampers the efficiency and power output of batteries. Additionally, manufacturing these materials at scale remains challenging and costly, delaying their widespread use.

ILs, another alternative, are salts in a liquid state at or near room temperature, comprising bulky organic cations and inorganic or organic anions. Their advantages include being non-volatile, non-flammable, and stable over a wide temperature range. ILs also exhibit high ionic conductivity, which is beneficial for fast-charging applications. However, these electrolytes face issues related to high production costs and complex synthesis processes, which have limited their adoption [15]. Despite these challenges, both SSEs and ILs have contributed valuable insights and paved the way for the development of the next generation of electrolytes [16].

One of the most promising developments in the search for greener, more efficient electrolytes is the emergence of DESs. DES remains in a liquid state at room temperature, which is beneficial for practical applications. Unlike conventional organic solvents, DES synthesized from renewable resources, aligning well with green chemistry principles. The synthesis process of DES is relatively simple and does not require complex or expensive equipment, which adds to their appeal as sustainable alternatives to traditional electrolytes [17].

What makes DESs particularly valuable for energy storage is their tunable nature. By adjusting the types and ratios of the HBDs and HBAs, researchers can customize the ionic conductivity, viscosity, and electrochemical stability of DES to suit specific applications. Figure 8.3 illustrates the diverse applications of deep eutectic solvents across multiple scientific and industrial domains. This adaptability is especially advantageous in battery design, where the electrolyte must meet stringent performance, safety, and environmental criteria. In lithium-ion and other battery systems, DESs offer wider electrochemical windows than many traditional electrolytes, allowing for higher operating voltages and increased energy density. Additionally, DESs are more thermally stable, which enhances battery safety and reduces risks associated with overheating [18]. Moreover, DES-based electrolytes improve the life cycle and efficiency of batteries, addressing one of the primary concerns with current battery technologies. Their low volatility and thermal stability contribute to fewer instances of degradation over time, which can extend battery lifespan. By mitigating issues like electrolyte decomposition and electrode corrosion, DES electrolytes reduce maintenance requirements and disposal concerns, aligning with goals for sustainable technology. In terms of environmental impact, DESs are advantageous because they can often be derived from bio-based materials, and their biodegradability ensures that they do not accumulate as harmful pollutants at the end of their life cycle [19].

The promising properties of DESs have opened doors to their use not only in traditional LIBs but also in next-generation battery chemistries, such as sodium-ion and magnesium-ion batteries, which are being explored as alternatives to lithium due to resource scarcity and cost concerns [37].

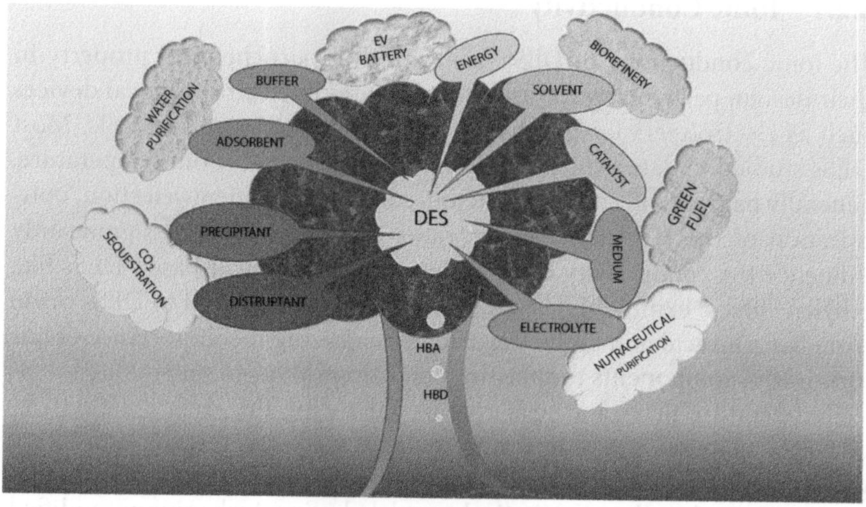

Figure 8.3 Diverse applications of deep eutectic solvents across fields.

DESs are also being studied for use in supercapacitors and other energy storage systems that require rapid charge-discharge cycles and high-power densities. As research into DESs continues, they represent a significant step forward in the quest for high-performing, environmentally friendly electrolytes, supporting the broader transition toward sustainable energy solutions [38]. In summary, the evolution of green electrolytes, from aqueous solutions to DESs, highlights a trajectory of innovation focused on achieving safe, efficient, and eco-friendly energy storage technologies.

8.2 Physicochemical and Electrochemical Behavior of DESs

Each type of DESs has distinct physicochemical properties, driven by the balance of forces between the components, influencing their solubility, ionic conductivity, and ability to dissolve various compounds. This classification highlights the diverse ways in which DESs can be engineered for specific industrial and scientific applications based on these properties. In particular, DESs possess several characteristics that make them suitable for adaptation as green electrolytes, including high ionic conductivity, efficient ion transport, a wide electrochemical potential window, low volatility, non-flammability, thermal and chemical stability, and biodegradability [41].

8.2.1 Ionic Conductivity

The ionic conductivity of DESs is a critical physicochemical property in their design, particularly for use as electrolytes in electrochemical devices such as electrolyzers for green hydrogen production and fuel cells. Most DESs studied to date exhibit low ionic conductivity at room temperature, generally below 2 mS cm^{-1}. Key factors such as component selection, optimization of the HBA:HBD molar ratio, and water content significantly influence the conductivity of DESs. For instance, polyol-based DESs like ethylene glycol exhibit higher ionic conductivity compared to DESs made from benzilic acid and diglycolic acid, emphasizing the importance of tailoring DES components to specific applications [49].

Temperature plays a crucial role in enhancing the ionic conductivity of DESs, as increasing temperature reduces viscosity, thereby promoting greater ionic mobility. This temperature-dependent behavior mirrors that of viscosity and can be quantitatively modeled using Arrhenius or modified

Arrhenius equations (e.g., the Vogel–Fulcher–Tammann, or VFT, model). These equations describe the inverse relationship between viscosity and ionic conductivity, with increasing temperature facilitating enhanced ionic transport. The relationship between DES conductivity and viscosity can be further explored through Walden plots, which offer qualitative insights into the ionicity of DESs. A Walden plot, a log-log plot of molar conductivity versus the inverse of viscosity, provides a relative measure of ionicity compared to standard references like KCl solutions. DESs with lower viscosity and high ionicity generally exhibit superior conductivity, whereas those with stronger HBD-HBA interactions and higher ionicity often show reduced conductivity. Understanding and optimizing these properties is essential for designing DESs with tailored ionic conductivity for energy storage and conversion technologies [43].

8.2.2 Ion Transport

Ion transport in DESs is crucial for the performance of batteries, as it directly influences ionic conductivity and cycle stability. The use of DESs as electrolytes significantly enhances ion transport compared to traditional organic electrolytes. DESs, composed of an HBD and an HBA, form a highly stable yet flexible network of hydrogen bonds. This dynamic network facilitates the dissociation of salts, leading to higher ionic conductivity. Additionally, the low volatility and high polarity of DESs promote a uniform ionic environment, reducing resistance and enhancing the mobility of charge carriers. Unlike organic electrolytes, DESs exhibit superior solvation of lithium or other cationic species, minimizing ion pairing and improving the efficiency of ion transport. Furthermore, the tunability of DESs allows for the optimization of viscosity and conductivity to balance electrochemical performance and stability, making them ideal candidates for next-generation energy storage systems like LIBs and redox flow batteries [44].

In LIBs, DESs synthesized from ChCl and urea exhibit an ionic conductivity that is significantly higher than that of traditional organic electrolytes, thereby enhancing the overall performance. Here, the solid electrolyte interphase (SEI), formed between the electrolyte and the electrode surface, plays a significant role in ion conduction. A stable SEI allows lithium ions (Li^+) to pass through effectively, maintaining low internal resistance and improving ionic conductivity. In contrast, a thick or unstable SEI (>50 nm) can increase resistance, reducing conductivity by 10%–20%. For instance, a stable SEI can maintain double the capacity retention after 500 cycles compared to unstable SEIs. The formation of the SEI in DESs is influenced by their unique hydrogen-bonding interactions, with DESs like ChCl and urea forming thinner and more stable SEIs compared to organic solvents, which enhances both ionic conductivity and cycle stability. Thus, optimizing DESs to promote a stable and efficient SEI is essential for improving battery performance over extended cycles [45].

8.2.3 Wide Electrochemical Window

The electrochemical stability window (ESW) is a critical property of an electrolyte, defining the voltage range within which it can operate without decomposing. For DESs, this property is particularly significant because their wide ESW enables them to support high-energy applications, such as LIBs, where operational voltages often exceed 4 V. The ESW of DESs is determined by the choice of HBDs and HBAs, as well as by the presence of additional salts or additives. This property is influenced not only by the intrinsic characteristics of the HBD and HBA but also by the interactions between the DES and the electrode surfaces. A wider ESW minimizes the risk of side reactions, such as electrolyte oxidation at the cathode or reduction at the anode, which can lead to performance degradation and safety concerns. Optimizing the HBD-HBA pairing allows researchers to tailor DESs to meet the specific voltage requirements of various battery chemistries [46].

Investigations have shown that DESs can exhibit a wide ESW depending on their composition and the salts added. A ChCl-based DES paired with ethylene glycol as the HBD displayed an ESW 1.5-fold lower than DESs formed with more electrochemically stable components, such as lithium salts or aromatic compounds. A DES composed of ChCl and urea, with the addition of lithium bis(trifluoromethanesulfonyl)imide (LiTFSI), demonstrated an ESW twice as high as that of regular DESs, enabling its use in high-voltage battery systems [47].

8.2.4 Low Volatility and Non-Flammability

Traditional organic electrolytes, commonly used in LIBs, pose significant safety risks due to their high volatility, low flash points (typically below 50°C), and flammability. These properties can lead to thermal runaway, leakage, or even explosions under mechanical or thermal stress. Safety is a critical consideration in battery technology, where high energy density and reliable performance are required under challenging conditions. DESs offer an inherently safer alternative due to their low volatility, high thermal stability, and non-flammable nature. Their vapor pressures are several orders of magnitude lower than that of organic solvents like ethylene carbonate, significantly reducing the risks of evaporation and leakage, even at elevated temperatures. Additionally, DESs exhibit remarkable thermal stability, with decomposition temperatures often exceeding 250°C. For example, a DES based on ChCl and glycerol has been reported to remain stable up to 280°C, greatly enhancing the safety profile of batteries in high-temperature environments. Another critical advantage of DESs is their non-flammability. Unlike conventional solvents with flash points below −20°C, DESs generally lack measurable flash points due to their negligible volatility and ionic nature. This property makes them highly suitable for applications where external impacts, overcharging, or high temperatures could trigger fire hazards. Furthermore, their high specific heat capacity and thermal conductivity contribute to better heat dissipation, reducing the likelihood of localized overheating. These safety features make DESs ideal for EVs and other high-performance energy storage systems, which require operation across a broad temperature range (−30°C to 60°C) while ensuring durability and minimizing risks. With their unique combination of stability, safety, and performance, DESs represent a transformative solution to the safety challenges of next-generation batteries [48].

8.2.5 Thermal and Chemical Stability

The thermal and chemical stability of DESs is a pivotal attribute that sets them apart as electrolytes for high-performance energy storage systems. DESs demonstrate exceptional thermal stability, with decomposition temperatures typically ranging between **200°C and 300°C**, depending on the choice of HBDs and HBAs. For instance, DESs formed from ChCl and glycerol or ethylene glycol have been shown to remain thermally stable up to **250°C**, ensuring reliable performance in environments with fluctuating or high temperatures, such as those encountered in EVs. The high thermal stability of DESs not only minimizes the risk of decomposition but also maintains their ionic conductivity and viscosity over a wide temperature range. Investigations reported that DESs retain a stable ionic conductivity (~**5–8 mS/cm**) even when operating at elevated temperatures of **80°C–100°C**, which is crucial for sustaining efficient ion transport in LIBs under demanding conditions.

In terms of chemical stability, DESs exhibit low reactivity with commonly used electrode materials, such as lithium metal or transition metal oxides. This stability reduces the formation of undesirable SEIs or side reactions, which can, otherwise, degrade the electrodes and compromise battery efficiency. Experimental studies have shown that DESs have ESWs ranging from **3.5** to **4.5 V**, which aligns well with the operational voltage requirements of LIBs [49]. The reduced tendency of DESs to react with lithium or other active materials is attributed to their unique ionic structure and strong hydrogen bonding, which shields reactive species. Additionally, DESs can be further optimized with additives to enhance their compatibility with specific electrode materials, extending battery lifespan and ensuring consistent performance over prolonged usage cycles.

8.2.6 Biodegradability and Biocompatability

DESs are gaining traction for their environmental sustainability, offering significant advantages over traditional organic electrolytes that are often toxic, non-biodegradable, and challenging to dispose of. DESs are typically composed of biodegradable and biocompatible components such as ChCl, urea, organic acids, and sugars. Studies have shown that many DESs exhibit low toxicity levels classifying them as practically toxic-free according to the globally harmonized system (GHS) of classification. The biodegradability of DESs has also been confirmed in studies where natural DES components such as sugars and organic acids degrade within **28 days**, with over **60% mineralization** under aerobic conditions, satisfying organisation for economic co-operation and development (OECD) standards for biodegradability. This aligns well with sustainability goals, making DESs safer for the environment at the end of their life cycle [50].

In addition to their green chemistry profile, many DESs are derived from renewable resources, such as plant-based ChCl and biomass-derived organic acids. For instance, glycerol, commonly used as a HBD in DESs, is a by-product of biodiesel production, further supporting circular economy initiatives. The carbon footprint of DES production is also notably lower, with estimated reductions of **20%–30%** compared to conventional organic solvents due to simplified synthesis and lower energy requirements. The non-hazardous nature of DESs also reduces risks during manufacturing and usage. Unlike volatile organic solvents, DESs have negligible vapor pressures (<**0.01 Pa** at room temperature), minimizing exposure risks for workers and the surrounding environment. This also eliminates significant hazards associated with flammable or volatile electrolyte systems, making DESs safer to handle and dispose of in large-scale applications [51].

8.2.7 Tunability

DESs are distinguished by their inherent tunability, allowing their properties to be customized for a wide range of electrochemical applications.

By modifying the choice of HBD and HBA, as well as their molar ratios, researchers can optimize properties such as ionic conductivity, viscosity, and thermal stability. For instance, the ionic conductivity of DESs can range from 10^{-5} to 10^{-3} S/cm at room temperature, depending on their composition. A ChCl–ethylene glycol DES exhibits twice the conductivity of a ChCl-urea DES, which can be further enhanced by adding lithium salts, increasing the conductivity up to fourfold [52]. Similarly, DES viscosity varies widely, typically between 10 cP and 1,000 cP, depending on the formulation. For example, a ChCl–ethylene glycol DES has significantly lower viscosity than a ChCl-glycerol DES, and the inclusion of co-solvents such as water can further reduce the viscosity by up to 50%. This tunability is highly beneficial for electrochemical applications. DESs specifically tailored for LIBs enable stable cycling for over 1,000 cycles with Coulombic efficiencies exceeding 99.5%. In sodium-ion and redox flow batteries, specific modifications to DESs enhance energy density and ionic exchange efficiency by up to 15% compared to conventional systems [53].

8.2.8 Viscosity

One of the main challenges for DESs compared to conventional electrolytes is their relatively high viscosity, which can hinder ion mobility and slow down ionic transport. The viscosity of DESs, depending on their formulation, is significantly higher than traditional organic solvents like carbonate-based electrolytes, which often have viscosities below 10 cP. For example, a ChCl-urea DES has a viscosity of approximately 750 cP, whereas ChCl–ethylene glycol DES has a lower viscosity of about 35 cP. Such high viscosities can reduce ionic conductivity, which generally lies between 10^{-5} and 10^{-3} S/cm for DESs, compared to organic electrolytes, where conductivities typically exceed 10^{-2} S/cm [54]. To mitigate this issue, researchers have explored strategies such as incorporating co-solvents like water or organic additives, which can significantly reduce viscosity without compromising the electrochemical stability. For instance, adding 10% (by weight) water to a ChCl-urea DES can lower viscosity by over 50%, leading to a two- to threefold increase in ionic conductivity. Similarly, selecting alternative HBD and HBA pairs with inherently lower viscosities can enhance ion mobility; a ChCl–ethylene glycol DES exhibits improved conductivity (~3.5 mS/cm) due to its relatively low viscosity.

Despite the viscosity challenges, DESs compensate with enhanced mechanical stability and lower volatility, reducing risks of leakage and evaporation. These attributes make them suitable for long-term applications, where mechanical durability and safety are critical. The trade-offs

Table 8.1 Physicochemical studies on deep eutectic solvents.

HBA	HBD	Molar ratio HBA:HBD	Viscosity μ_0 (mPa·s)	Density (g/cm^3)	Conductivity γ_0(mS cm^{-1})	Surface tension (mN m^{-1})	References
MTPB	EG	1:3.5	2.7726×10^{-4}	-	664.5	-	[20]
Thymol	oleA	1:1	23.4	0.913	0.4	31.6	[21]
TBAC	oleA	1:1	460.6	0.911	636	32.9	[21]
ChCl	EG	1:2	36	1.12	7.61	-	[22]
Chcl	Pyruvic acid	1:2	-	-	-	-	[23]
ChCl	Phenol	1:2	99.8	1.09	1.4	-	[24]
ZNH	Gly	1:3	720	-	3.32	-	[25]
ZNH	Gly	1:6	920	-	1.39	-	[25]
ZC	LA	1:1	-	1.325	0.0046	-	[26]
ChCl*	EG*	1:2	33.1	-	17.4	-	[27]
DEAC	EG	3:1	-	1.101	-	-	[28]
DEAC	EG	2:1	-	1.099	-	-	[29]
TBAB	Diethanoa Mine	1:1	-	1.0651	-	-	[30]

(Continued)

Table 8.1 Physicochemical studies on deep eutectic solvents. (*Continued*)

HBA	HBD	Molar ratio HBA:HBD	Viscosity μ_0 (mPa·s)	Density (g/cm^3)	Conductivity γ_0(mS cm^{-1})	Surface tension (mN m^{-1})	References
TBAB	EG	3:1	-	1.139	-	-	[31]
TBAC	Glutamic acid	1:10	-	0.9612	-	-	[32]
ChCl	Levulinic acid	2:1	171.3	-	-	-	[33]
ChCl	Malonic acid	1:1	1,389	-	-	-	[34]
ChCl	Glucose	1:1	1,809	-	-	-	[35]
MTPB	2,2,2-Trifluoroacetamide	8:1	136.5	-	-	-	[36]

between viscosity and conductivity continue to be a focus of research, with innovations aimed at striking a balance to maximize DES performance in advanced battery systems [55].

8.2.9 Density

The density is a critical physicochemical property, influencing the formulation of equations of state that enable the prediction of additional thermodynamic characteristics. At room temperature, most DESs exhibit densities greater than water (>1 g/cm^3). The differences in DES densities can be explained using hole theory, which suggests that DESs consist of void spaces similar to molten salts and ILs, with ionic mobility restricted by the average size of these voids. Consequently, incorporating larger molecular components into a DES typically results in higher density. For instance, zinc chloride–based DESs demonstrate significantly higher densities across all temperature ranges compared to commonly studied DESs like reline and glyceline, attributed to reduced void space radii. The density of DESs is also affected by the characteristics of their HBA and HBD components, including alkyl chain length, functional group count, and molar ratio. DESs containing carboxylic acids, for example, tend to exhibit higher densities than glycerol-based systems. A notable study found that increasing the zinc chloride molar ratio to 3 raised the DES density to over 2 g/cm^3 at room temperature [56]. Table 8.1 presents an overview of physicochemical studies conducted on DESs prepared at various molar ratios and analyzed at room temperature.

8.3 Advantage of DES as Electrolytes

The Kamlet–Taft parameters—hydrogen-bond–donating ability (α), hydrogen-bond–accepting ability (β), and dipolarity/polarizability (π*)—provide a quantitative framework to understand the solvent behavior of DESs. These parameters directly impact the performance of DESs as electrolytes in energy storage applications like LIBs [57]. The hydrogen-bond–donating parameter, **α**, represents the strength of hydrogen bonding contributed by the HBD in the DES. For example, ChCl combined with ethylene glycol exhibits an α value of approximately **0.76**, whereas, with urea as the HBD, α increases to **1.10**, indicating stronger hydrogen-bond–donating behavior. A higher α facilitates efficient anion solvation, promoting the dissociation of salts like LiTFSI or $LiPF_6$, which enhances the ionic conductivity of the DES. This is crucial for enabling charge transport within the battery. The hydrogen-bond–accepting parameter, β, measures the ability of the DES to accept hydrogen bonds, which is significant for cation solvation. ChCl exhibits a **β** value of around **0.86** when combined with ethylene glycol, and this increases to **0.94** with glycerol. Such β values ensure effective stabilization of lithium ions (Li^+) in the electrolyte, reducing ion pairing and aggregation. This property supports smoother ionic transport and helps mitigate issues such as dendrite formation, leading to improved battery cycle stability. The dipolarity/polarizability parameter, π*, evaluates the ability of DESs to stabilize ionic species through dipole-dipole interactions. For instance, DESs formed from ChCl and urea exhibit a π* value of approximately **0.90**, indicating strong polarizability. This high π* value supports the stabilization of both cations and anions, ensuring efficient ionic conduction. However, a trade-off exists: excessive π* values can increase the viscosity of the DES, as observed in some DESs where viscosity exceeds **100 cP**, leading to reduced ion mobility. Thus, optimizing π* within a range of **0.8 to 1.2** is essential for balancing ionic conductivity and viscosity. The combination of these parameters allows the design of DESs with tailored properties for specific electrochemical applications. For instance, in LIBs, a DES with $\alpha > $ **1.0**, **β** $\approx$ **0.9**, and π* in the range of **0.9 to 1.0** could achieve an ionic conductivity of **10^{-3} to 10^{-2} S/cm** at 25°C, which is suitable for practical applications [58]. By leveraging Kamlet–Taft parameters, researchers can systematically engineer DESs for high ionic conductivity, stability, and efficiency, addressing the challenges of electrolyte design for sustainable energy storage technologies.

8.3.1 Computational Validation of DES Battery Electrolytes

Computational models are indispensable for optimizing electrolyte formulations and predicting their behavior under varying conditions. Figure 8.4 illustrates different computational approaches used for designing DESs. Molecular dynamics (MD) simulations provide detailed insights into atomic and molecular interactions in electrolytes, enabling the analysis of ion transport, solvation structures, and stability. For instance, MD studies of DESs reveal ionic conductivity, diffusion coefficients of lithium ions, and viscosities, which vary on the basis of specific DES compositions and temperature. These parameters are crucial for optimizing electrolyte performance [59].

Complementing MD, density functional theory (DFT) offers quantum-level insights into electronic structures and reactivity. For DESs, DFT-calculated highest occupied molecular orbital (HOMO) energies typically

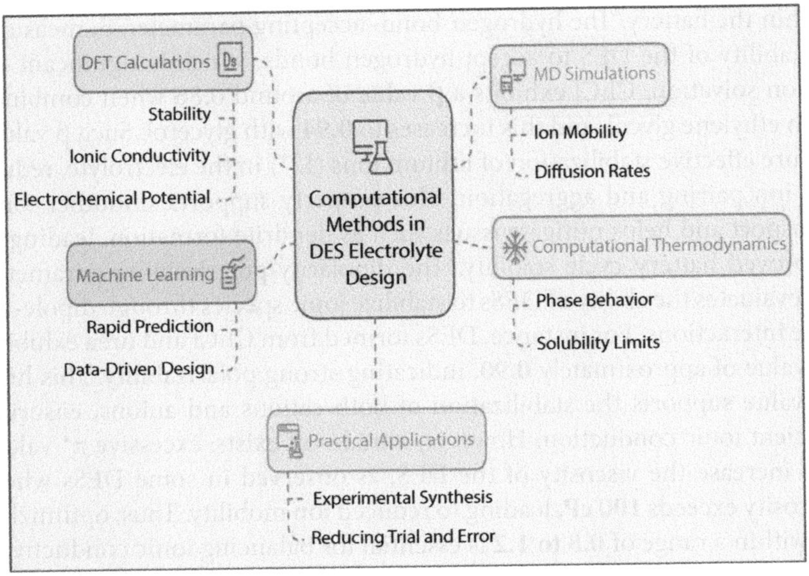

Figure 8.4 Computational approaches for designing deep eutectic solvents.

range from −6 to −5.5 eV, whereas lowest unoccupied molecular orbital (LUMO) energies fall between −2.5 and −1.8 eV, resulting in a HOMO-LUMO gap of approximately 3 to 4 eV. This gap is essential for ensuring oxidative and reductive stability within the operational voltage range of LIBs, which spans 2.5 to 4.2 V. Additionally, DFT simulations provide solvation energies for lithium ions in DESs, correlating with efficient ion transport and dendrite suppression. The ESWs of DESs, derived from DFT, typically range from 3.5 to 5 V, ensuring compatibility with LIB requirements. Furthermore, DFT predicts hydrogen bond interactions and their binding energies, which significantly influence viscosity and ionic conductivity [60]. DFT studies also reveal the formation of SEIs with low energy barriers for decomposition, supporting stable and long-lasting LIB performance. These findings underscore the tunable nature of DESs and their potential as sustainable, high-performance electrolytes.

Continuum models are critical for understanding the macroscopic behavior of electrolytes, focusing on bulk properties and large-scale interactions. These models are built on the Poisson–Nernst–Planck (PNP) equations, combining the Poisson equation for electrostatics and the Nernst–Planck equation for ion transport. Using these models, researchers can predict critical parameters such as ion concentration profiles, electric fields, and current densities. Continuum models also highlight the influence of temperature, showing that increased temperatures improve ionic conductivity and reduce resistance. Moreover, these models reveal the formation of depletion layers near electrodes due to ion consumption, significantly affecting lithium-ion transport and deposition. Such findings are pivotal for optimizing electrolyte formulations and battery designs to minimize overpotential, mitigate dendrite growth, and enhance safety and efficiency in energy storage systems [61].

Quantum mechanical/molecular mechanical (QM/MM) methods, the Lattice–Boltzmann method (LBM), and Monte Carlo (MC) simulations are essential computational tools for studying electrolyte behavior in complex systems. QM/MM approaches excel at investigating localized phenomena, such as reaction mechanisms at electrode-electrolyte interfaces. In QM regions, quantum mechanics accurately calculates electronic properties, while the MM framework describes the surrounding environment. For example, QM/MM studies of SEI formation in LIBs have revealed detailed reaction pathways and energy barriers, typically ranging from 0.5 to 2.5 eV, depending on the electrolyte composition and applied potentials. These insights help optimize electrolyte additives and stabilize the SEI layer. The LBM is particularly effective for modeling fluid flow and ion transport in porous media, making it invaluable for studying porous electrodes. LBM simulations predict electrolyte flow velocities of 0.1 to 1.0 mm/s in porous battery separators under typical operating conditions, with pressure drops quantified at 10^4 to 10^5 Pa across the electrode thickness. These results aid in designing porous structures that minimize resistance and enhance ion transport, crucial for high-performance batteries [62].

MC simulations rely on statistical sampling to explore thermodynamic properties and equilibrium states. MC approaches calculate solubility limits and predict phase behaviors in electrolyte systems. They have also been used to determine equilibrium constants and Gibbs free energy changes (ΔG) for dissolution and ion-pairing reactions [63]. Together, these computational methods provide a comprehensive toolkit for probing electrolyte properties, from atomistic reactions to macroscopic transport. This facilitates the development of advanced energy storage systems with optimized electrolytes and electrode architectures.

Computational approaches such as computational fluid dynamics (CFD), thermodynamic models, machine learning (ML), and phase-field simulations also play pivotal roles in advancing battery electrolyte research. CFD models simulate electrolyte flow and ion transport within systems like flow batteries, providing velocity fields, pressure drops, and temperature distributions critical for optimizing battery designs.

8.3.2 Role of DES in Batteries and Supercapacitors

Deep eutectic electrolytes (DEEs) have gained attention for their potential to outperform traditional electrolytes in batteries and supercapacitors due to their unique properties, such as tunable viscosity, ionic conductivity, and electrochemical stability.

8.3.3 Key Electrical Properties of DES-Based Electrolytes

The enhanced electrical characteristics of DESs as electrolytes in LIBs make them an attractive alternative to conventional organic solvents. Energy density refers to the amount of energy a battery can store per unit of mass or volume. The energy density of DES-based LIBs is typically in the range of 150 to 200 Wh/kg; this is lower than that of conventional electrolytes. Despite this, DESs can still be suitable for applications where moderate energy storage is sufficient, such as in some EVs or portable devices. Energy density can be further optimized by enhancing the electrochemical stability and improving ion transport through the electrolyte system [42, 64].

Power density, expressed in W/kg, is another important parameter indicating how quickly energy can be delivered by the battery. For DES-based batteries, the power density typically ranges from 100 to 500 W/kg, which is appropriate for moderate-power applications. Power density measures the rate at which a battery can release its stored energy, and it is especially crucial for applications requiring rapid bursts of energy. Although DES electrolytes may not match the higher power densities of traditional organic electrolytes, their ability to stabilize the SEI layer and facilitate efficient ion transport contributes to their competitiveness in moderate-power applications [42, 65]. The C-rate refers to the rate at which a battery is charged or discharged relative to its capacity. A 1C rate implies that the battery is charged or discharged in one hour, whereas higher C-rates such as 2C or 5C indicate faster charging or discharging. DES-based LIBs generally perform well at moderate C-rates, typically up to 5C, without significant degradation. A C-rate of 5C would imply that the battery is charged or discharged in 12 minutes, which is suitable for many high-performance applications requiring fast charging or discharging. The performance at higher C-rates, however, can be influenced by the viscosity and stability of the electrolyte, with increased internal resistance potentially leading to a drop in efficiency [66].

Table 8.2 Performance enhancement of deep eutectic solvents with varied HBA-HBD combinations as electrolytes in energy storage systems.

HBA	HBD	Molar ratio	Results	References
ChCl	Urea	1:3 (water:DES)	Viscosity reduced by 75%; peak current density increased by 5–9.5 times; ionic conductivity increased by 5–10 times; ohmic resistance decreased by 60%–90%	[42]
Terpene	TOPO	Terpene:TOP O with varying LiTFSI concentrations	Dissolves up to 30% molar LiTFSI; **ionic conductivity:** low, ranging from 3.6×10^{-3} to 9.3×10^{-2} mS cm^{-1} at 25°C; **diffusion coefficients:** similar for Li^+ and TFSI–, suggesting long-lived ion pair formation	[40]
Fluorinated polymer	UPy	--	Self-healing capability, wide electrochemical window, high migration number, and superior interface stability due to fluorinated groups and eutectic nature	[70]
Cross-linked PVA	SSA, Na^+	PVA:SSA functionalized with Na groups	**Ionic conductivity:** tunable, over 2.910 mS cm^{-1}; **hydrogen bond dynamics (HBDy):** polarization impedance of 52.87 kΩ; **Na^+ diffusion:** polarization impedance of 12.90 kΩ	[71]

(*Continued*)

Table 8.2 Performance enhancement of deep eutectic solvents with varied HBA-HBD combinations as electrolytes in energy storage systems. (*Continued*)

HBA	HBD	Molar ratio	Results	References
DMA TFE	Hydrated aluminum perchlorate	Al perchlorate: DMA:TFE = specific ratio in ADT45	Low-cost, intrinsically safe, and stable; **solid electrolyte interface (SEI):** AlF3-rich SEI layer suppresses side reactions	[72]
$Mg(TFSI)_2$	Urea		Formation of anion-rich urea-Mg^{2+} coordination structures; enhancing Mg^{2+} migration in the solid-state electrolyte	[73]
ChCl	(1,2-BD)		Highest ionic conductivity (~10^{-8} S cm^{-1})	[74]
$ZnCl_2$	Acetamide (C_2H_5NO), H_2O		Excellent thermal stability of the PAA-DES GPE; inhibition of Zn dendrite growth	[39]
DES-based electrolyte	Bis (3-amino-propyl)–terminated PDMS (NH_2); fluoroethylene carbonate (FEC)		High ionic conductivity (0.60 mS cm^{-1} at 30°C), wide electrochemical voltage window (4.5 V vs. Li/Li^+), specific capacity of 151.6 mAh g^{-1}	[75]
ZNH	EG	1:6	Highest ionic conductivity (σ): 7.51 mS/cm; wider electrochemical potential window (EPW): 1.06 V, indicating better ion delocalization and reversibility	[76]
Imidazole (ImL)	LiTFSI	4:1	Ionic conductivity: 4.5×10^{-3} S cm^{-1} at 80°C	[46]

(*Continued*)

Table 8.2 Performance enhancement of deep eutectic solvents with varied HBA-HBD combinations as electrolytes in energy storage systems. (*Continued*)

HBA	HBD	Molar ratio	Results	References
Pyrazole (PyL)	LiTFSI	4:1	Ionic conductivity: 5.8 × 10^{-3} S cm^{-1} at 80°C; initial specific capacity of 147 mAh g^{-1}	[46]
Sodium perchlorate zinc chloride	Acetamide		Broad electrochemical stability window and self-extinguishing behavior; improved coulombic efficiency; discharge capacity: 76.4 mAh g^{-1} at 0.3 A g^{-1}	[77]
1,2-Dimethylimidazole (DMIm)	LiTFSI		Ionic conductivity: 1.8 × 10^{-4} S cm^{-1} at 60°C Li+ transference number: 0.41	[78]
ChCl	Gly	1:2	Electrochemical stability up to 2.5 V; low viscosity	[79]
ChCl	PA	1:2	Specific capacitance of 145 F g^{-1}	[79]
BMIM[$MeSO_3$]	NMAc, formamide		Energy density of 16.3 Wh kg^{-1}; viscosity: <20 cp (lower than pure ILs); ionic conductivity: >5 mS/cm (higher than pure ILs)	[80]

Coulombic efficiency is a measure of how efficiently a battery charges and discharges, calculated as the ratio of charge extracted from the battery to the charge put into it, expressed as a percentage. In DES-based LIBs, Coulombic efficiency typically exceeds 98%, which is comparable to or even better than traditional organic electrolytes. High Coulombic efficiency indicates that most of the charge inserted into the battery is recoverable during discharge, making the battery highly efficient over extended cycles. This high efficiency is partly attributed to the formation of a stable SEI layer, which prevents loss of active lithium ions during cycling [67, 68].

The cycle life of a battery refers to the number of charge-discharge cycles it can undergo before its capacity degrades significantly (usually by 20%). DES-based electrolytes have demonstrated impressive cycle life performance, often ranging from 500 to 1,000 cycles with 90% capacity retention. This durability is essential for long-term applications, such as in EVs, where long cycle life is crucial. The stable SEI formation facilitated by DESs helps maintain battery performance over many cycles, preventing the issues commonly encountered with conventional organic electrolytes, such as dendrite growth or rapid SEI degradation [69]. Table 8.2 provides a detailed analysis of the enhanced performance outcomes for various combinations of HBAs and HBDs in DESs utilized as electrolytes in energy storage systems.

8.4 DES-Based Eutectogels in Energy Storage

Eutectogels, derived from DESs combined with gel matrices, represent an innovative approach to creating SSEs for batteries. These materials merge the ionic conductivity of DESs with the mechanical stability of gels, making them suitable for electrochemical applications like lithium-ion and redox flow batteries. Their tunable physicochemical properties, eco-friendliness, and cost-effectiveness further enhance their appeal for sustainable energy storage solutions [81]. One of the key advantages of DES-based eutectogels is their high ionic conductivity. For example, a eutectogel comprising ChCl and urea (in a 1:2 molar ratio) embedded in a poly(vinyl alcohol) (PVA) matrix has demonstrated a conductivity of **1.2×10^{-3} S/cm**. Additionally, these materials exhibit wide ESWs, often exceeding **4.0 V**. A eutectogel formed with ChCl:ethylene glycol DES achieved an ESW of **4.4 V**, suitable for high-voltage battery systems. The mechanical strength of DES eutectogels typically falls within **0.1 to 1 MPa**, providing robust yet flexible membranes that can withstand operational stresses. Furthermore, their thermal stability enables operation across a broad temperature range, from **−20°C to 100°C**, allowing for use in extreme environmental conditions.

The use of DES-based eutectogels in batteries significantly enhances performance. For instance, these materials mitigate dendrite growth in lithium-metal batteries by facilitating the formation of stable SEIs. A eutectogel composed of ChCl and glycerol in a PVA framework exhibited minimal dendrite formation over **100 charge-discharge cycles**. Additionally, batteries incorporating DES eutectogels have shown impressive cycle life, retaining **98% of their capacity after 200 cycles** at a 1C rate. The superior ion transport properties of eutectogels contribute to higher energy densities, with some designs achieving **200 Wh/kg**, comparable to commercial organic electrolyte systems [82].

Despite their advantages, DES-based eutectogels face challenges, such as limited mechanical strength and compatibility issues with certain electrode materials. Addressing these challenges requires the development of advanced polymer matrices or the incorporation of nanomaterials to reinforce the eutectogel structure. Additionally, optimizing DES formulations for specific battery chemistries and scaling up production processes will be essential for their commercial viability.

By combining enhanced ionic conductivity, wide ESWs, excellent thermal stability, and eco-friendly characteristics, DES-based eutectogels are poised to play a transformative role in next-generation energy storage systems. They offer a compelling synergy of performance, sustainability, and adaptability, aligning with the demands of modern battery technologies [83].

The enhanced performance of DEEs in batteries and supercapacitors is driven by several factors. First, DEEs typically offer comparable or even superior ionic conductivity compared to traditional organic or IL electrolytes, which improves the charge and discharge rates in both types of energy storage devices. Additionally, DEEs often have a wider ESW, allowing for higher voltage operation, which directly translates to increased energy density in supercapacitors and batteries. This makes DEEs suitable for devices where high-voltage operation is critical [72].

Another key advantage is the lower viscosity seen in certain DEE mixtures, such as ChCl with ethylene glycol. This reduced viscosity allows for better ion mobility, improving the overall rate performance of both batteries and supercapacitors. DEEs also contribute to the safety and eco-friendliness of energy storage systems, as they are less flammable and toxic than conventional organic electrolytes, making them a safer choice for large-scale applications.

In terms of durability, DEEs can significantly improve the cycle life of batteries. For example, sodium-ion and lithium-sulfur batteries using DEEs have shown enhanced cycle retention and reduced capacity fading over extended usage, addressing common challenges associated with traditional electrolytes. In sodium-ion batteries, for instance, DEEs provide a 20% improvement in cycle life, whereas lithium-sulfur batteries show a 30% enhancement in stability. Additionally, the use of DEEs reduces issues like dendrite formation in zinc-ion batteries, further extending cycle life and reliability. Overall, DEEs deliver multiple performance benefits, such as improved ionic conductivity, broader electrochemical stability, enhanced cycle life, and better safety profiles, making them an excellent choice for next-generation batteries and supercapacitors [38, 84].

8.5 Challenges and Future Prospects

DESs face several challenges that must be addressed for their successful application in battery technologies. A significant issue is their relatively high viscosity that can hinder ion mobility and subsequently reduce battery efficiency. Ongoing research aims to develop low-viscosity DES formulations or explore strategies to mitigate viscosity-related problems. Another challenge lies in optimizing DES compositions for specific battery systems. Despite the tunable properties of DESs, determining the best combinations for various battery chemistries necessitates further investigation. This includes a deeper understanding of how DESs interact with different electrode materials, which is essential for maximizing overall performance.

Additionally, transitioning DESs from laboratory success to commercial viability poses considerable hurdles. Although DESs demonstrate promise in controlled environments, scaling up production for industrial use remains a significant challenge. It is crucial to assess the cost-effectiveness of DESs, ensuring that they are economically viable compared to traditional electrolytes. Moreover, their long-term stability and compatibility with existing manufacturing processes must be thoroughly evaluated to facilitate integration into commercial battery systems. Addressing these challenges will be key to realizing the full potential of DESs in advanced battery applications and achieving widespread adoption in the energy storage market.

8.6 Conclusion

DEEs represent a major advancement in the development of greener, safer, and more efficient energy storage systems. These electrolytes, made from biodegradable and abundant substances like choline chloride, urea, and glycerol, offer high ionic conductivity, thermal stability, eco-friendly, and tunability, making them ideal candidates for next-generation batteries and supercapacitors. Research has shown that specific combinations, such as ChCl-urea, ChCl–ethylene glycol, and ChCl-glycerol, can significantly improve performance in terms of capacitance, energy density, and cycle stability. DEEs have also emerged as a viable eco-friendly alternative to conventional electrolytes, alongside aqueous electrolytes and ILs derived from renewable sources. Their simple preparation process and ability to address safety and environmental concerns position them as key players in the future of sustainable energy technologies. Further optimization and research are expected to unlock even more potential in enhancing the performance and longevity of energy storage devices.

AI disclosure statement: The author acknowledges the use of AI-based tools for minor language editing or grammatical correction. However, all intellectual contributions, analyses, and conclusions presented in this chapter are entirely the author's own.

References

1. Kang, Z., Ye, Z., Lam, C.-M., Hsu, S.-C., Sustainable electric vehicle charging coordination: Balancing CO2 emission reduction and peak power demand shaving. *Appl. Energy*, 349, 121637, 2023.
2. Vengatesan, S., Jayakumar, A., Sadasivuni, K.K., FCEV vs. BEV — A short overview on identifying the key contributors to affordable & clean energy (SDG-7). *Energy Strat. Rev.*, 53, 101380, 2024.
3. Ping, P., Luo, X., Gao, W., Li, Z., Kong, D., Electrochemical and thermal properties of hard carbon anodes in ether-based and ester-based electrolytes for sodium ion batteries. *J. Energy Storage*, 107, 114921, 2025.
4. Viswanathan, A. and Shetty, A.N., High energy and rate capable supercapacitor of polyaniline/vanadium pentoxide nanocomposite and its green electrolyte. *Next Sustain.*, 5, 100088, 2025.
5. Sharma, R., Sharma, A., Thakur, R.C., Effect of Deep Eutectic Solvent (DES) on physicochemical and electrochemical properties of triethylmethylammonium tetrafluoroborate in propylene carbonate and aqueous propylene carbonate solvent systems for energy storage applications: A comparative analysis. *J. Mol. Liq.*, 416, 126494, 2024.
6. Singh, M., Kaur, S., Mehta, S., Kumar, M., Kumar, K., Meena, S.K., *et al.*, Boosted Capacity and Stability of Aqueous Iron-Sulfur Battery using DMSO as an Electrolyte Additive. *Energy Storage Mater.*, 75, 103965, 2024.
7. Mandai, K., Miao, S., Dowaki, K., Environmental impact assessment on polymer electrolyte fuel cell co-generation system, lithium-ion battery, and photovoltaic hybrid system combination and operation, considering performance degradation. *Clean. Eng. Technol.*, 20, 100756, 2024.
8. Sahoo, J.P., Ray, R., Prakash, R., Sivasubramani, S., Impacts of Plug-in EVs and decentralized power generation on distribution system operation. *e-Prime Adv. Electr. Eng. Electron. Energy*, 9, 100658, 2024.

9. Wei, M., Zhai, P., Li, Y., Zhao, X., Li, J., Zhang, T., *et al.*, Flexible ionic liquid aided "LAGP in PVDF" quasi- solid-state electrolyte for high performance and stable Li metal batteries. *Int. J. Hydrogen Energy*, 79, 1278–88, 2024.
10. Jamalpour, S., Maghsoudi, R., Azizi, A., Tailored graft polymerization on SiO2 nanoparticle by sulfonate styrene and methyl methacrylate and evaluation of their electrochemical performance in gel polymer electrolyte based on poly (vinylidene fluoride) for Li-ion batteries application. *Electrochem. Commun.*, 166, 107779, 2024.
11. Ünlü, B., Türk, S., Özacar, M., Gellan gum/PEDOT:PSS gel electrolyte and application on quasi-solid dye sensitized solar cells. *J. Photochem. Photobiol., A*, 450, 115471, 2024.
12. Qi, Y. and Xu, M., Engineering towards stable sodium metal anodes in room temperature sodium- sulfur batteries: challenges, progress and perspectives. *Energy Storage Mater.*, 72, 103704, 2024.
13. Gao, Y., Fan, M., Cheng, X., Liu, X., Yang, H., Ma, W., *et al.*, Deep eutectic solvent: Synthesis, classification, properties and application in macromolecular substances. *Int. J. Biol. Macromol.*, 278, 134593, 2024.
14. Shuaibu, A.D., Shah, S.S., Alzahrani, A.S., Aziz, M.A., Advancing gel polymer electrolytes for next-generation high-performance solid-state supercapacitors: A comprehensive review. *J. Energy Storage*, 107, 114851, 2025.
15. Zhang, N., Zhao, X., Liu, G., Peng, Z., Wu, J., Men, M., *et al.*, Solid electrolyte membranes for all-solid- state rechargeable batteries. *eTransportation*, 20, 100319, 2024.
16. Safavi-Mirmahalleh, S.-A., Dadkhah-Janqor, A., Salami-Kalajahi, M., Chitosan-based gel polymer electrolytes for high performance Li-ion battery. *Int. J. Biol. Macromol.*, 282, 137304, 2024.
17. Ye, L., Zhao, Z., E-x, L., Wang, L., One-pot two-step synthesis of γ-keto sulfones in deep eutectic solvent. *Mendeleev Commun.*, 34, 5, 709–10, 2024.
18. Ramalingam, L.P., Ramalingam, B., Rathnasamy, S., Kathirvelu, P., Deep eutectic solvent-based sustainable electrochemical lithium batteries – Prospects, challenges, and life cycle engineering. *Sustain. Energy Technol. Assess.*, 73, 104136, 2025.
19. Millia, L., Dall'Asta, V., Ferrara, C., Berbenni, V., Quartarone, E., Perna, F.M., *et al.*, Bio-inspired choline chloride-based deep eutectic solvents as electrolytes for lithium-ion batteries. *Solid State Ionics*, 323, 44–8, 2018.
20. Hayyan, M., Hayyan, A., Hafizi, A.D.M., Basirun, W.J., Yeow, A.T.H., Salleh, M.Z.M., *et al.*, Phosphonium- based deep eutectic solvents: Physicochemical properties and application in Zn-air battery. *Chem. Eng. Process. - Process Intensif.*, 185, 109310, 2023.
21. Duque, A., Sanjuan, A., Bou-Ali, M.M., Alonso, R.M., Campanero, M.A., Physicochemical characterization of hydrophobic type III and type V deep eutectic solvents based on carboxylic acids. *J. Mol. Liq.*, 392, 123431, 2023.
22. Smith, E.L., Abbott, A.P., Ryder, K.S., Deep Eutectic Solvents (DESs) and Their Applications. *Chem. Rev.*, 114, 21, 11060–82, 2014.

23. Chenthamara, B. and Gardas, R.L., Unravelling the influence of water: Investigating the physicochemical properties of deep eutectic solvents based on choline chloride and pyruvic acid/α- keto glutaric acid. *J. Mol. Liq.*, 416, 126489, 2024.
24. Kityk, A., Pavlik, V., Hnatko, M., Exploring deep eutectic solvents for the electrochemical and chemical synthesis of photo- and electrocatalysts for hydrogen evolution. *Int. J. Hydrogen Energy*, 48, 100, 39823–53, 2023.
25. Getie, F.A., Ayele, D.W., Habtu, N.G., Yihun, F.A., Yemata, T.A., Ambaw, M.D., *et al.*, Binary glycerol-based deep eutectic solvents containing zinc nitrate hexahydrate salt for rechargeable zinc air batteries applications with enhanced properties. *Heliyon*, 9, 7, e17810, 2023.
26. Sarjuna, K. and Ilangeswaran, D., Preparation of some zinc chloride based deep eutectic solvents and their characterization. *Mater. Today: Proc.*, 33, 2767–70, 2020.
27. Sun, P., Lu, P., Xu, J., Ma, Q., Zhang, W., Shah, A.A., *et al.*, The influence and control of ultrasonic on the transport and electrochemical properties of redox couple ions in deep eutectic solvent (DES) for redox flow battery application. *Electrochim. Acta*, 394, 139140, 2021.
28. Ibrahim, R.K., Hayyan, M., AlSaadi, M.A., Ibrahim, S., Hayyan, A., Hashim, M.A., Physical properties of ethylene glycol-based deep eutectic solvents. *J. Mol. Liq.*, 276, 794–800, 2019.
29. Siongco, K.R., Leron, R.B., Li, M.-H., Densities, refractive indices, and viscosities of N,N- diethylethanol ammonium chloride–glycerol or –ethylene glycol deep eutectic solvents and their aqueous solutions. *J. Chem. Thermodyn.*, 65, 65–72, 2013.
30. Murshid, G., Mjalli, F.S., Naser, J., Al-Zakwani, S., Hayyan, A., Novel diethanolamine based deep eutectic mixtures for carbon dioxide (CO2) capture: synthesis and characterisation. *Phys. Chem. Liq.*, 57, 4, 473–90, 2019.
31. Sedghamiz, M.A. and Raeissi, S., Physical properties of deep eutectic solvents formed by the sodium halide salts and ethylene glycol, and their mixtures with water. *J. Mol. Liq.*, 269, 694–702, 2018.
32. Mjalli, F.S., Novel amino acids based ionic liquids analogues: Acidic and basic amino acids. *J. Taiwan Inst. Chem. Eng.*, 61, 64–74, 2016.
33. Ullah, R., Atilhan, M., Anaya, B., Khraisheh, M., García, G., ElKhattat, A., *et al.*, A detailed study of cholinium chloride and levulinic acid deep eutectic solvent system for CO 2 capture *via* experimental and molecular simulation approaches. *Phys. Chem. Chem. Phys.*, 17, 32, 20941–60, 2015.
34. Florindo, C., Oliveira, F.S., Rebelo, L.P.N., Fernandes, A.M., Marrucho, I.M., Insights into the Synthesis and Properties of Deep Eutectic Solvents Based on Cholinium Chloride and Carboxylic Acids. *ACS Sustain. Chem. Eng.*, 2, 10, 2416–25, 2014.
35. Aroso, I.M., Paiva, A., Reis, R.L., Duarte, A.R.C., Natural deep eutectic solvents from choline chloride and betaine – Physicochemical properties. *J. Mol. Liq.*, 241, 654–61, 2017.

36. Kareem, M.A., Mjalli, F.S., Hashim, M.A., AlNashef, I.M., Phosphonium-Based Ionic Liquids Analogues and Their Physical Properties. *J. Chem. Eng. Data*, 55, 11, 4632–7, 2010.
37. Xue, W., Li, M., Zhao, Y., Deng, Y., Li, Y., Direct electrochemical reduction of solid Bi2O3 to Bi as a high-performance anode for sodium-ion batteries using deep eutectic solvent. *J. Energy Storage*, 98, 113062, 2024.
38. Samage, A., Kuppe, P., Halakarni, M., Ganesan, B.K., Kamath, S.V., Yoon, H., *et al.*, Room temperature and rapid synthesis of ZnMn2O4 nanostructured spinel using deep eutectic solvent for high energy asymmetric supercapacitors. *J. Energy Storage*, 97, 112934, 2024.
39. Liu, X., Wang, J., Lv, P., Zhang, Y., Li, J., Wei, Q., Gel polymer electrolyte based on deep eutectic solvent in flexible Zn-air batteries enables dendrite-free Zn anode. *Energy Storage Mater.*, 69, 103382, 2024.
40. de Araujo Lima e Souza, G., Pelegano-Titmuss, E., Muñoz, M., Gurkan, B., di Pietro, M.E., Mele, A., *et al.*, Probing the potential of type V Deep eutectic solvents as sustainable electrolytes. *J. Mol. Liq.*, 416, 126526, 2024.
41. Saleem, I., Sheikh, A., Khan, A.Y., Ahmed, S., Abbas, M., Physicochemical properties of ammonium acetate/propionic acid as a novel deep eutectic solvent and its binary mixtures with water in (298.15– 353.15) K range. *Spectrochim. Acta, Part A*, 313, 124144, 2024.
42. Wang, W., Mu, A., Wang, Y., Wang, J., Yang, B., Iron-vanadium redox flow batteries electrolytes: performance enhancement of aqueous deep eutectic solvent electrolytes *via* water-tuning. *Electrochim. Acta*, 504, 144926, 2024.
43. Koyakkat, M., Moriyama, K., Asakura, S., Kawamoto, H., Shirota, H., Deep eutectic solvents based on ammonium iodide and iodine possessing high electrical conductivity. *J. Mol. Liq.*, 384, 122250, 2023.
44. Liu, Z., Li, H., Yao, H., Zhuang, Y., Gao, R., Wang, Z., *et al.*, A PEGylated deep eutectic solvent for "bubbling" synthesis of SnO2/SnS heterostructure for the stable lithium-ion storage. *J. Colloid Interface Sci.*, 682, 995–1005, 2025.
45. Wang, X., Zhou, J., Wang, H., Zhang, J., Ma, X., Qiu, S., *et al.*, Mildly expanded graphite with exceptional performance from waste lithium ion batteries by space-confined intercalation of deep eutectic solvent. *Sep. Purif. Technol.*, 354, 129329, 2025.
46. Liu, R.-L., Yang, W., Wang, Q., Du, X., Hu, B., Chen, Y., *et al.*, Azole derived deep eutectic electrolyte for high performance lithium metal-free full batteries. *Ceram. Int.*, 50, 5, 8178–84, 2024.
47. Moradi, H. and Farzi, N., The investigation of deep eutectic electrolyte based on Choline Chloride: Ethylene glycol in 1:3 M ratio and lithium hexafluorophosphate salt for application in Lithium-Ion batteries. *J. Mol. Liq.*, 360, 119476, 2022.
48. Chidiac, J., Timperman, L., Anouti, M., Physical properties and compatibility with graphite and lithium metal anodes of non-flammable deep eutectic

solvent as a safe electrolyte for high temperature Li-ion batteries. *Electrochim. Acta*, 408, 139944, 2022.
49. Xu, Q., Qin, L.Y., Ji, Y.N., Leung, P.K., Su, H.N., Qiao, F., *et al.*, A deep eutectic solvent (DES) electrolyte- based vanadium-iron redox flow battery enabling higher specific capacity and improved thermal stability. *Electrochim. Acta*, 293, 426–31, 2019.
50. Wu, H., Pang, Z., Ji, L., Pang, X., Li, Y., Yu, X., Strong adhesive, high conductive and environmentally stable eutectogel based on "Water in deep eutectic Solvent": Ultrasensitive flexible temperature and strain sensors. *Chem. Eng. J.*, 497, 154883, 2024.
51. Hayyan, A., Zainal-Abidin, M.H., Putra, S.S.S., Alanazi, Y.M., Saleh, J., Nor, M.R.M., *et al.*, Evaluation of biodegradability, toxicity and ecotoxicity of organic acid-based deep eutectic solvents. *Sci. Total Environ.*, 948, 174758, 2024.
52. Cheng, X., Xuan, T., Wang, L., A low-cost all-iron hybrid redox flow batteries enabled by deep eutectic solvents. *Chem. Eng. J.*, 491, 151936, 2024.
53. Lim, E., Yon Pang, J.-S., Lau, E.V., A comparative study of thermophysical properties between choline chloride-based deep eutectic solvents and imidazolium-based ionic liquids. *J. Mol. Liq.*, 395, 123895, 2024.
54. Lakshmi Papa Rao, G., Mandal, A., Pal, N., Choline chloride-Urea based deep eutectic Solvent: Characterization, interfacial behavior and Synergism in binary (surfactant) systems. *Chem. Phys.*, 588, 112496, 2025.
55. Makarov, D.M. and Kolker, A.M., Viscosity of deep eutectic solvents: Predictive modeling with experimental validation. *Fluid Phase Equilib.*, 587, 114217, 2025.
56. Peng, D. and Minceva, M., Predicting the density and viscosity of deep eutectic solvents at atmospheric and elevated pressures. *Fluid Phase Equilib.*, 582, 114086, 2024.
57. O'Mari, O. and Vullev, V.I., Optical Window to Polarity of Electrolyte Solutions. *Molecules (Basel Switzerland)*, 28, 11, 2023.
58. Tsurumaki, A., Trequattrini, F., Palumbo, O., Panero, S., Paolone, A., Navarra, M.A., The effect of ether- functionalisation in ionic liquids analysed by DFT calculation, infrared spectra, and Kamlet–Taft parameters. *Phys. Chem. Chem. Phys.*, 20, 12, 7989–97, 2018.
59. Behnajady, B., Seyf, J.Y., Karimi, S., Moradi, M., Sohrabi, M., Molecular dynamic (MD) simulation and density function theory (DFT) calculation relevant to green leaching of metals from spent lithium-ion battery cathode materials using glucose-based deep eutectic solvent (DES). *Hydrometallurgy*, 223, 106223, 2024.
60. Vorobyova, V., Skiba, M., Vinnichuk, K., Linyucheva, O., Vasyliev, G., Influence of hydrogen bond donor compound type on the properties of betaine based deep eutectic solvents: A computational and experimental approach. *J. Mol. Struct.*, 1323, 140743, 2025.

61. Prado, D.M., Shen, X., Savinell, R., Burda, C., Hydrodynamic voltammetry of Fe2+/3+ in aqueous deep eutectic solvents towards redox flow batteries. *Electrochim. Acta*, 467, 143082, 2023.
62. Ma, Q., Xu, Q., Chen, Q., Chen, Z., Su, H., Zhang, W., Lattice Boltzmann model for complex transfer behaviors in porous electrode of all copper redox flow battery with deep eutectic solvent electrolyte. *Appl. Therm. Eng.*, 160, 114015, 2019.
63. Han, M., Zhou, J., Fan, H.J., Opportunity for eutectic mixtures in metal-ion batteries. *Trends Chem.*, 5, 3, 214–24, 2023.
64. Huang, Y.-C., Chu, Y.-L., Chang, C.-C., Liu, L.-Y., Teng, H., Chen, B.-H., *et al.*, Ultraviolet-cured composite polymer electrolyte containing deep-eutectic-solvent and aluminum oxide filler for lithium-metal batteries with enhanced cycle life. *J. Electroanal. Chem.*, 940, 117455, 2023.
65. Wang, Y., Xu, R., Xiao, B., Lv, D., Peng, Y., Zheng, Y., *et al.*, A poly(1,3-dioxolane) based deep-eutectic polymer electrolyte for high performance ambient polymer lithium battery. *Mater. Today Phys.*, 22, 100620, 2022.
66. Bais, A., Subhedar, D., Panchal, S., Experimental investigations of a novel phase change material and nano enhanced phase change material based passive battery thermal management system for Li-ion battery discharged at a high C rate. *J. Energy Storage*, 103, 114395, 2024.
67. Cheng, L., Wang, Z., Wang, T., Wu, Y., Liu, X., Zhang, Y., Understanding and research progress on the initial coulombic efficiency of silicon-based anodes in lithium-ion batteries. *J. Electroanal. Chem.*, 973, 118670, 2024.
68. NP, S., Sreenath, S., Dave, V., Kumar, P., Verma, V., Nagarale, R.K., Investigation of modified deep eutectic solvent for high performance vanadium redox flow battery. *Electrochim. Acta*, 508, 145242, 2024.
69. Zhou, H., Yang, Y., Zhang, Z., Wang, W., Yang, L., Du, X., Charge and discharge strategies of lithium-ion battery based on electrochemical-mechanical-thermal coupling aging model. *J. Energy Storage*, 99, 113484, 2024.
70. Yu, Y., Ling, C., Yang, J., Zhu, A., Naren, T., Chen, Y., *et al.*, Self-Healing fluorinated polymer deep eutectic electrolytes for stable lithium metal batteries. *Chem. Eng. J.*, 498, 155376, 2024.
71. Halilu, A. and Hashim, M.A., Deconvoluting the distribution of relaxation times for charge transport descriptors in solid state deep eutectic electrolytes. *J. Power Sources*, 614, 235016, 2024.
72. Zhao, T., Wang, F., Li, Z., Cheng, R., Ran, B., Fu, J., *et al.*, Nonflammable localized high-concentration deep eutectic electrolytes for safe and stable rechargeable aluminum batteries. *Nano Energy*, 131, 110298, 2024.
73. Chen, F., Yu, J., Li, R., Shi, F., Che, X., Chan, K.C., *et al.*, Direct crystallization of deep eutectic solvent into solid-state electrolyte for magnesium metal batteries. *J. Power Sources*, 611, 234780, 2024.
74. Zaharuddin, I.Q., Zailani, N.A.M., Nazir, K., Ghani, F.S.A., Latif, F.A., Abidin, S.Z.Z., Exploring the role of deep eutectic solvents in enhancing the

properties of PMMA for potential electrolyte films. *Int. J. Electrochem. Sci.*, 19, 9, 100730, 2024.

75. Chen, Y., Ling, C., Long, K., Liu, X., Xiao, P., Yu, Y.-Z., *et al.*, A siloxane-based self-healing gel electrolyte with deep eutectic solvents for safe quasi-solid-state lithium metal batteries. *Chem. Eng. J.*, 488, 150888, 2024.
76. Getie, F.A., Ayele, D.W., Habtu, N.G., Yemata, T.A., Yihun, F.A., Ambaw, M.D., *et al.*, Development of zinc nitrate hexahydrate salt (ZNH)/ethylene glycol (EG) binary composite deep eutectic electrolytes. *Int. J. Electrochem. Sci.*, 19, 7, 100616, 2024.
77. Lin, T.-W., Kumar, M.S., Shen, H.-H., Lin, J.-Y., Acetamide-based deep eutectic solvents as efficient electrolytes for K–MnHCFe//Zn dual-ion batteries. *J. Power Sources*, 614, 234972, 2024.
78. Ye, W., Wang, J., Shi, Z., Guo, K., Xue, Z., Facile fabrication of cross-linked polymer electrolyte *via* imidazole-based deep eutectic solvent-induced *in situ* polymerizations. *J. Membr. Sci.*, 696, 122541, 2024.
79. Wang, W., Sabugaa, M.M., Chandra, S., Asmara, Y.P., Alreda, B.A., Ulloa, N., *et al.*, Choline chloride- based deep eutectic solvents as electrolytes for wide temperature range supercapacitors. *J. Energy Storage*, 71, 108141, 2023.
80. Choudhury, S., Mahanta, U., Prasanna Venkatesh, R., Banerjee, T., Ionic liquid derived novel deep eutectic solvents as low viscous electrolytes for energy storage. *J. Mol. Liq.*, 366, 120245, 2022.
81. Chuang, R.-M., Vo, T.H., Tsao, H.-K., Sheng, Y.-J., High-strength poly(vinyl alcohol) physical eutectogels: Effects of polymer molecular weight, DES composition, and heat treatment. *J. Mol. Liq.*, 417, 126592, 2025.
82. Thombre, A.V., Das, R., Kundu, D., Lithium ion transportation in lignin infused polyvinyl alcohol based eutectogel: A molecular dynamics framework. *J. Energy Storage*, 102, 114060, 2024.
83. Li, H., Chen, H., Kang, Q., Guo, L., Huang, X., Xu, H., Gel polymer electrolyte for flexible and stretchable lithium metal battery: Advances and prospects. *Chin. Chem. Lett.*, 36, 9, 110325, 2024.
84. Sun, H., Li, F., He, X., Li, Z., Chen, X., Wang, C., *et al.*, Largely improved cycling stability of CoV2O6 supercapacitor electrode synthesized by deep eutectic solvent. *J. Alloys Compd.*, 992, 174555, 2024.

9

Life Cycle Assessment—A Systemic Tool in Sustainability Management

Jayita Chopra

Department of Chemical Engineering, BITS Pilani K K Birla Goa Campus, Goa, India

Abstract

The accelerating pace of industrialization and the corresponding surge in global resource consumption have intensified environmental challenges, making sustainability a central concern for modern societies. Rising greenhouse gas concentrations, climate change, and resource depletion underscore the urgency of transitioning to more sustainable pathways for development. At the same time, the growing demand for food, fuel, water, and materials necessitates innovative strategies that balance human needs with ecological resilience. In response, the concepts of sustainable development, green chemistry, industrial ecology, and the circular economy have emerged as key frameworks guiding environmental policy and resource management. Within these frameworks, life cycle assessment (LCA) has become an indispensable tool, enabling systematic evaluation of the environmental impacts associated with products, processes, and technologies across their entire life cycles.

Initially introduced in the 1960s and later standardized by international bodies, LCA has evolved into a robust methodology for quantifying inputs such as raw materials, water, and energy, alongside outputs including products, emissions, and waste streams. By examining each stage of a product's life—from resource extraction and production to use and end-of-life disposal—LCA provides a comprehensive understanding of environmental footprints. This allows for the identification of both the strengths and limitations of emerging technologies, ensuring that decisions are based on holistic and unbiased assessments. Importantly, LCA not only highlights areas for improvement but also supports policymakers, industries, and researchers in evaluating the long-term sustainability of strategies such as renewable energy adoption, recycling, and resource recovery.

Email: jayitac@goa.bits-pilani.ac.in

Senthilkumar Rathnasamy, Vivek Rangarajan and Umile Gianfranco Spizzirri (eds.) Green Engineering and Sustainable Technology: Tools and Techniques, (249–264) © 2026 Scrivener Publishing LLC

In the context of the circular economy, where waste is increasingly viewed as a resource, LCA serves as a crucial instrument for determining the most environmentally friendly and economically viable methods for recycling, recovery, and reuse. Its systemic approach ensures that efforts aimed at reducing waste and promoting sustainability do not inadvertently shift environmental burdens to other stages of the product life cycle. By bridging the gap between technological innovation and environmental stewardship, LCA provides insights that are vital for achieving meaningful progress toward global sustainability goals.

This chapter explores the fundamental principles, methodologies, and applications of LCA. It outlines the different types of LCA studies, essential steps involved in conducting them, allocation procedures, and sensitivity analyses. Additionally, it discusses the limitations and prospects of LCA. Through this examination, the chapter highlights the critical role of LCA in advancing sustainable engineering practices and supporting informed decision-making for a more resilient future.

***Keywords*:** Life cycle assessment (LCA), environmental impact assessment, resource recovery and recycling, circular economy strategies, allocation procedures, sensitivity analysis

9.1 Introduction

In recent times, there has been a heightened societal consciousness regarding the trajectory of human progress and the consequential environmental impacts both now and in the future. The reason behind this can be attributed to the increased impact on the environment caused by rapid industrialization since the last century. The greenhouse gas responsible for global warming is present in the atmosphere at an all-time high since the beginning of humanity, causing a disastrous cascading effect on different components of the environment [12]. Moreover, with an increase in population and lifestyle demands, there has been a significant increase in demand for resources like fuel, fiber, fresh water, and food. Therefore, sustainable development and the somewhat less emphasized fields of green chemistry and industrial ecology have evolved into pivotal components of environmental policy and guiding principles for effective resource management. The substantial advancements in reducing fossil fuel dependency, coupled with the adoption of renewable energy sources, mark significant strides in the past few decades [5, 16]. Yet, the central concern remains: how sustainable are these renewable energy technologies, and how can we accurately quantify their environmental impacts? Addressing this pivotal question, life cycle assessment (LCA) has emerged as a potent tool, facilitating informed decisions about the sustainability of technologies and offering valuable insights for future directions [2, 4].

There are also efforts in the front of resource recovery and utilization, which is broadly termed as circular economy. The circular economy is a concept aimed at reducing waste production and generating revenue from the waste [13]. Efforts in this direction focus on recycling, waste minimization, and the sustainable use of resources to ensure long-term environmental health and economic viability. However, it is crucial to keep the technologies sustainable, and, therefore, LCA has become an indispensable tool in evaluating environmental impacts arising from recycling, recovery, and reuse of resources from waste products. By analyzing the life cycle of materials, LCA identifies the most sustainable methods for resource recovery, ensuring that the processes are environmentally friendly and economically viable [5, 6]. LCA was initially introduced in the 1960s. In 1969, The Coca-Cola company used LCA for the first time to compare and quantify the environmental footprints of different beverage bottle packaging (glass and Polyethylene Terephthalate (PET)) [3]. This was the first study that used the currently practiced method of inventory analysis considering the raw materials and fuels used. LCA gained prominence during the 1980s when solid waste management became a pressing issue. Over the years, different industrial sectors have been using LCA, and it has evolved with significant improvements in methodology and impact assessment methods [8]. In 1990, the international body of the Society of Environmental Toxicology and Chemistry (SETAC) organized a workshop titled "A Technical Framework for Life Cycle Assessment," which initiated a series of subsequent workshops to discuss the important elements of the LCA methodology. This series culminated in the Code of Practice workshop, which laid the groundwork for the first official guidelines for LCA. These guidelines were formalized and standardized in 1996 by the International Organization for Standardization (ISO) through four ISO standards: 14040 (principles and framework), 14041 (goal and scope definition), 14042 (life cycle impact assessment), and 14043 (life cycle interpretation). In 2006, the standards were consolidated into ISO 14040 and ISO 14044. According to ISO 14040, an LCA study involves evaluating the environmental impacts of a product's life cycle, including acquiring raw materials, production process, use, and disposal [25]. A process flow diagram for a complete LCA study has been demonstrated in Figure 9.1.

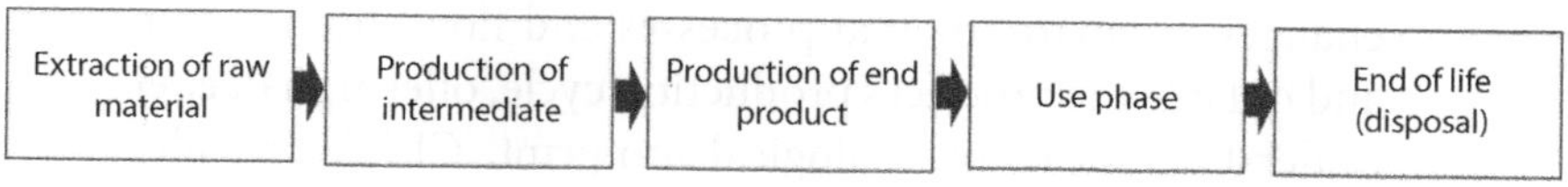

Figure 9.1 Simplified life cycle of a product [21].

Therefore, LCA involves accounting for all the inputs such as raw materials, water, and energy and outputs such as products, emissions, and waste streams for all the stages of the cycle for a product or a process. It enables a systematic assessment of the environmental footprints involved with the process. It facilitates the quantification of consumptions, emissions, and their impacts on endpoint categories [11]. LCA clarifies the advantages and disadvantages of the chosen technology and helps make unbiased decisions for the future. LCA has become almost synonymous with sustainable engineering. Conducting an LCA is essential for evaluating the environmental impacts associated with technologies before they reach the commercial market. This comprehensive assessment serves as a crucial tool for policymakers, enabling them to make informed decisions that promote sustainability. It helps policymakers understand the long-term implications of resource use, recycling, and waste management strategies. Additionally, LCA helps identify key processes that require improvement to reduce environmental footprints.

Therefore, this chapter aims to discuss the various types of LCA studies available, the steps involved, and their significance in a typical LCA study. It will cover allocation procedures, sensitivity analysis, and the limitations and prospects of LCA. Through this detailed examination, readers will gain a deeper understanding of how LCA contributes to sustainable engineering and environmental stewardship.

9.2 Types of LCA

There are two primary types of LCA methodologies:

i) Attributional LCA (ALCA), which focuses on quantifying the direct environmental impacts of a process or a product. This entails evaluating resources consumed and emissions generated throughout its life cycle, such as feed usage, transportation fuel, and heating electricity in biofuel production. ALCA determines the portion of global environmental burdens attributed to a product.

ii) Consequential LCA (CLCA), a more comprehensive approach that extends beyond direct impacts to include broader environmental repercussions. CLCA assesses changes in environmental processes and flows both within and outside the product's production cycle, offering a deeper understanding of its ecological footprint. CLCA provides insight into how the production and utilization of the product influence global environmental burdens.

In simpler words, ALCA tells what part of the global environmental burden is contributed by the product, whereas CLCA tells what is the impact of the product on the global environmental consequential impact of a decision. This differentiation emerged to address debates surrounding data inputs and allocation challenges inherent in LCAs. ALCA relies on average data and allocates environmental burdens by distributing them across the life cycles served by a process. Conversely, CLCA ideally employs marginal data throughout the life cycle and sidesteps allocation issues through system expansion. While CLCA offers greater accuracy, ALCA boasts other advantages. The choice between ALCA and CLCA should ideally be made by the LCA practitioner following discussions with the client and potentially involving other stakeholders and colleagues. Advanced CLCA can include the possibility of using a new energy technology instead of conventional energy sources and that can significantly impact the results.

In terms of realistic modeling, CLCA performs realistically, whereas ALCA relies on conceptual rules that deal with the splitting of co-product processes. In CLCA, any action is judged on the basis of its consequences or impacts, making it undoubtedly the preferred choice of LCA aiming to meet sustainable goals. Although ALCA has garnered significant attention regarding standards, it is essential to recognize that CLCA stands out as the method currently demonstrating consistent reliability in execution. This distinction arises primarily from the limitations within the existing life cycle impact assessment (LCIA) methodologies utilized in ALCA, which struggle to adequately partition environmental multi-input processes. Hence, advocating for a shift toward prioritizing CLCA within standards is imperative.

9.3 Application of LCA in Green Processes

The concept of sustainability has become indispensable in diverse production operations and processes, prompting environmental scientists to look at the environmental impacts resulting from the entire life cycle of a process or a product. Therefore, the need for a holistic and systematic tool gave rise to LCA, which avoids delusional perspectives of environmental burdens. To determine whether an aluminum car is truly more environmentally friendly than one made of steel, it is essential to consider all factors. Although using aluminum reduces gasoline consumption due to its lighter weight, the energy required to produce aluminum is significantly higher than that for steel. A comprehensive judgment can be made only by evaluating these and other relevant details [15]. LCA studies help guide future policies by determining the consequences of a product's life cycle or process. The decision is primarily based on how sustainable the process is.

Future technologies can be developed by building on existing LCA study results and incorporating necessary changes to ensure that they are validated, ethical, and aligned with humanitarian principles. When executed effectively, these processes can serve as powerful instruments for fostering a shared commitment to social responsibility and sustainability.

Many research studies involving LCA have been carried out in diverse fields, analyzing the environmental footprints of the process. A study of green methanol production from CO_2 has indicated that up to 1.75 tons of CO_2 can be abated per ton of methanol produced if renewable energy is used during production.

Another study [9] of techno-environmental assessment of green biorefinery processes concluded that higher environmental benefits can be obtained by increasing product yield rather than reducing energy consumption. Replacement of the conventional production process with biorefining of green biomass significantly improves the environmental burdens [10]. An LCA study on the application of biochar as green sorbent revealed that biochar as a soil ameliorant shows potential for global warming potential (GWP) mitigation and carbon sequestration and also mentions certain challenges, including variability in biochar properties [20]. A recent study on biofuel production in a biorefinery model highlights the environmental benefits when an integrated approach is adopted than the conventional process [7].

However, although LCA has been adopted for green chemistry and sustainable engineering for the production routes and processes, an integrated approach allowing more comprehensive analysis is essential.

9.4 Stages of LCA

LCA is based on four main phases: i) goal and scope definition, ii) inventory analysis, iii) impact assessment, and iv) interpretation. An LCA framework is shown in Figure 9.2.

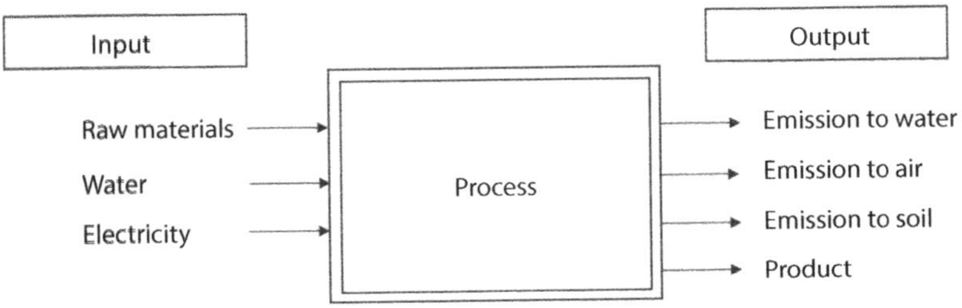

Figure 9.2 Inventory analysis for a product or process.

Goal and Scope: The goal and scope phase defines the objectives, outlining the intended application, motivations behind the research, and target audience. This entails critical methodological decisions, such as specifying the functional unit (FU), delineating system boundaries, determining allocation procedures, selecting impact categories for analysis, and identifying the LCIA models to be employed.

The significance of defining system boundaries has been underscored by several studies [18, 24]. The paramount boundary exists between the technological system and the natural environment. Ideally, all inputs and outputs essential to the product's function should be meticulously traced upstream and downstream, encompassing flows of energy or matter: (a) from nature to the technological system and (b) from the technological system back to nature. System boundaries necessitate specification across various dimensions:

- Distinctions between the technological system and its natural environment
- Geographical considerations
- Temporal constraints
- Capital goods production
- Delineations between the life cycle of the focal product and the interconnected life cycles of other products [23].

System boundary identification involves establishing criteria to determine which unit processes are included in a product system (ISO 14044 2006b). The delineation of system boundaries should align with the specific objectives of the life cycle study. Typically, four primary options exist for defining these boundaries: cradle to gate, gate to gate, cradle to grave, and gate to grave.

Inventory Analysis: During the life cycle inventory (LCI) phase, data collection and calculation procedures are employed to quantify the inputs and outputs of the system under study. The LCI initiates the acquisition of raw materials or energy carriers. For non-renewable resources like deposits, the cycle commences with extraction or prospecting. Fund resources, such as farmland, forests, and animals, involve activities from harvesting preparation like plowing, planting, fertilizing, and pesticide use to produce the harvest. Flowing resources like solar radiation and running water necessitate activities to integrate them into the technological system. The inventory supplies all inputs and outputs from the system calculated based on the chosen FU (Figure 9.3). Sometimes lab-scale data are extrapolated as per the FU to have meaningful life cycle analysis.

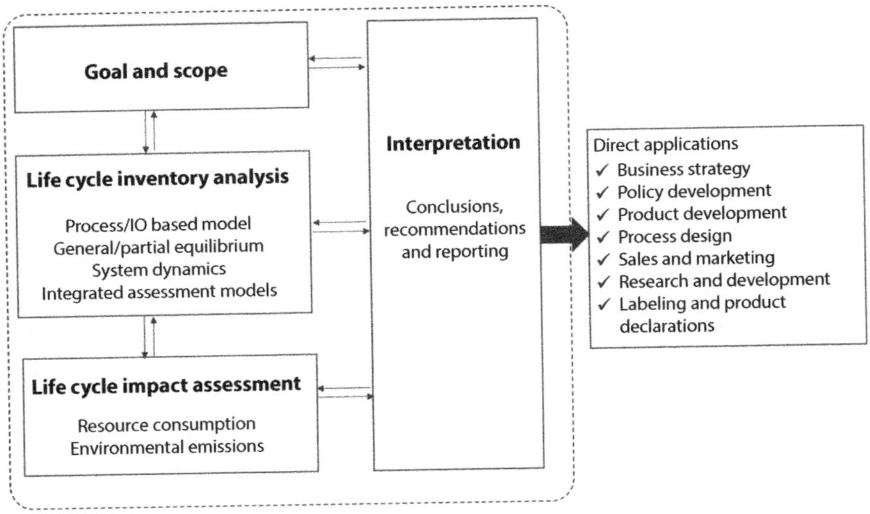

Figure 9.3 Stages of LCA [25].

Impact Assessment: LCA considers the various inputs and outputs that interact with the environment and quantifies the consequential environmental impacts. The impact assessment derives the potential impacts from these data. The LCIA aggregates the inputs and outputs with proper mass balance, emissions, resources, and energy use. It is defined as the stage of LCA that quantifies the different environmental impacts and the significance of the impact. It also helps to identify the raw material, unit process, or operation that contributes maximally to the overall impact, thereby serving as a future improvement guideline [1]. The distribution of impacts among the different sub-processes can be analyzed. For example, if chloroform is found to have more impacts, then it should be replaced in future studies with a greener alternative like ethanol or methanol.

Interpretation: The interpretation phase involves concluding the results of the impact assessment. Because all the impact categories have different units, it is not possible to compare the results arising from the different sub-processes. Normalization of the impact categories helps in understanding and interpreting the results effectively. Recommendations are then made on the basis of the study's objectives. The interpretation phase should produce results that align with the study objectives, reach clear conclusions, elucidate any limitations, and offer actionable recommendations. Interpretation is composed of the identification of critical issues followed by checking for completeness, sensitivity, and consistency.

9.5 Methodologies

There are many different characterization models available assessing the midpoint and endpoint indicators. There is no single LCA method that can be universally applicable in all scenarios. LCA methodologies are inherently goal- and scope-dependent. For instance, LCIA includes both midpoint-oriented and endpoint-oriented approaches. Some LCIA methods translate emissions, including gaseous, wastewater, and hazardous

Table 9.1 LCA characterization methods and impact categories.

Method	Impact category	
	Midpoint category	**Damage category**
ReCiPe midpoint (H)	Global warming Stratospheric ozone depletion Ionizing radiation Ozone formation Fine particulate matter formation Ozone formation Terrestrial acidification Freshwater eutrophication Marine eutrophication Terrestrial ecotoxicity Freshwater ecotoxicity Marine ecotoxicity Human carcinogenic toxicity Human non-carcinogenic toxicity Land use Mineral resource scarcity Fossil resource scarcity Water consumption	Human toxicity Ecosystem damage Resource depletion
TRACI	Ozone depletion Global warming Smog Acidification Eutrophication Carcinogenic Non-carcinogenic Respiratory effects Ecotoxicity Fossil fuel depletion	

substances, into impact category indicators at the midpoint level, such as acidification, climate change, and ecotoxicity. Others combine the impact of these midpoint categories at the endpoint level, like damage to human health, resources, and ecosystem quality [14]. The development of the ReCiPe 2008 methodology is one of the recent advancements in LCIA. Other commonly used methodologies are IMPACT 2002+, Eco-indicator 99, EPS 2000, and TRACI (Tool for the Reduction and Assessment of Chemical and Other Environmental Impacts), developed by the U.S. Environmental Protection Agency (EPA). Table 9.1 shows the list of commonly used LCA methods and the respective impact categories.

9.6 Allocation

Allocation is crucial for distributing environmental impacts among multiple products during their production, recycling, and disposal phases. The primary challenge lies in fairly distributing these environmental burdens. This distribution is typically based on various factors such as the amount produced, calorific value, molecular mass, electricity equivalent, or economic value. For instance, calorific value is often used for allocation in cases where products have substantial differences in their energy content, such as in refinery and biofuel products. The choice of allocation method significantly influences the results of an LCA study, making careful and relevant allocation essential for achieving transparent and accurate outcomes [19].

9.7 Uncertainty and Sensitivity Analysis

The goal of a sensitivity analysis is to estimate the uncertainties in LCA results arising from factors such as raw data, system boundaries, allocation, and the choice of impact categories. This process typically involves examining scenarios that vary in the modeling of the product system, with each scenario focusing on a single parameter under investigation. Sensitivity analysis descriptively determines and documents the influence of the altered parameter on the final results. To conduct a robust analysis, it is essential to identify the most crucial parameter and explore different scenarios by varying it. For instance, in an energy-intensive process, scenarios could involve changing the electricity source from fossil fuels to solar or hydroelectric power, which would likely result in significant variations in environmental impacts. Similarly, scenarios can be developed on the basis of variations in product productivity, examining cases of both low and high productivity to understand their effects on the overall impacts.

9.8 Case Study on LCA

Case Study I: *Biofuel Production from Beach-Cast Seaweed by Pyrolysis and Hydrothermal Liquefaction* [17]

Goal and Scope Definition: The goal of the study was to select the most environmentally friendly and carbon-neutral thermal technology for bio-oil production from beach-cast macroalgae.
System Boundary: Only the preparation of biomass and production of biofuel was included; transportation was excluded. System boundary for both pyrolysis and HTL has been shown in Figure 9.4.
Functional Unit: 1 ton of fuel produced.
Inventory: The seaweed biomass had a moisture content of 84%. The pyrolysis gas and bio-oil yield were 23% and 32%, respectively. For hydro-thermal liquefaction (HTL), bio-oil yield of 16.6%, a gas phase yield of 16%, and a coal yield of 40% were considered.
Sensitivity Analysis: The analysis was performed using two important parameters—electricity and *Ulva* sp. moisture content.
Life Cycle Impact Assessment: Beach-cast seaweed biomass (*Ulva* sp.) was used as the feedstock for producing heavy fuel oil (HFO) by pyrolysis and HTL. The base case scenario was the production of HFO from fossil

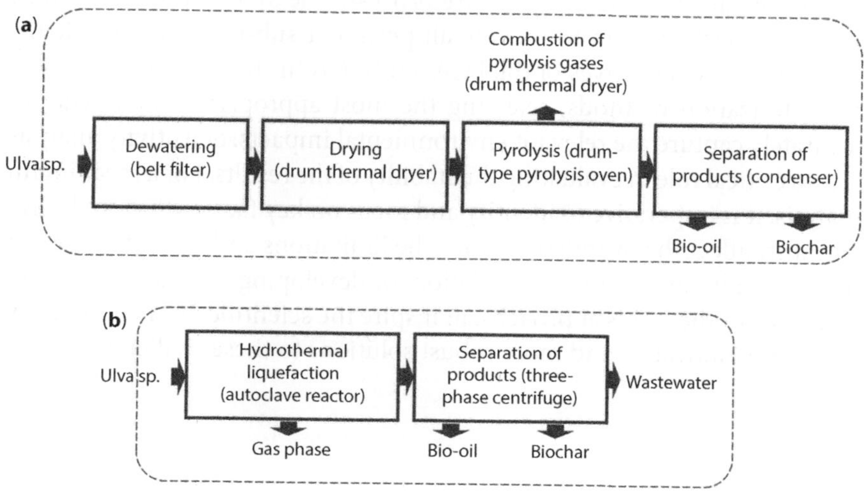

Figure 9.4 System boundary of (a) pyrolysis and (b) HTL.

fuel sources. Pyrolysis showed a GWP100 of 884.3 kg CO_2-Eq per 1 Mg of fuel; however, HTL would have a higher GWP100 of 969.6 kg CO_2-Eq per 1 Mg of fuel. Shifting energy sources from traditional to wind-based improved the GWP100, which is calculated as 838 and 628 kg CO_2-Eq per 1 Mg of fuel for pyrolysis and HTL, respectively. Sensitivity analysis results suggested an optimal case where electricity is obtained from a wind turbine (1–3 MW) with an optimal moisture content of biomass to be 70%; however, it did not show much influence on the results.

9.9 Limitations and the Way Forward

There are several inherent limitations in each stage of an LCA study. No single rule can be universally applied to all LCA studies, and it is often impossible to encompass every environmental aspect across all stages of a process. Accessing original raw data can be challenging, leading to a reliance on numerous assumptions that introduce variability into the interpretations. LCA is a simulation tool, and the results are determined by the assumptions and system boundaries. The accuracy of an LCA study is closely tied to the precision and availability of the primary raw data, and the chosen system boundary can significantly influence the results. Additionally, comparing the environmental impacts of two similar processes is difficult due to potential differences in FUs, system boundaries, characterization methods, assumptions, and allocations [22].

The goal and scope should be defined by an individual with comprehensive technical expertise in the entire process. The system boundary must be meticulously chosen to include all pertinent sub-processes and faithfully reflect the objectives of the LCA study. Given the variety of available characterization methods, selecting the most appropriate one is vital to accurately capture the relevant environmental impacts. Sensitivity analysis plays a critical role in evaluating the stability of the results. For a meaningful analysis, it is imperative to identify and focus on key factors that could significantly affect the overall outcome. The limitations and gaps identified in an LCA study can serve as a foundation for developing a research roadmap for future studies. This approach can inspire the scientific LCA community to explore innovative and more robust solutions to these challenges.

9.10 Conclusion

LCA emerges as a key systemic tool used in all technology development domains, ranging from pharmaceutical production to waste-to-energy, automotive, food industry, biofuel, and many more. LCA serves as a decision-making tool for policymaking, product development, and marketing, and improving the environmental footprint of a process. LCA methodology evaluates and analyzes the impacts of all processes within a given system, encompassing the entire life cycle while accounting for local constraints and specific conditions. The particular challenges associated with each step of LCA analysis have been discussed in the chapter and the possible recommendations have been provided. A techno-economic study will complement the LCA analysis providing information on the economic feasibility of the process.

AI disclosure statement: The author acknowledges the use of AI-based tools for minor language editing or grammatical correction. However, all intellectual contributions, analyses, and conclusions presented in this chapter are entirely the author's own.

References

1. Bakshi, B.R., Life Cycle Impact Assessment, in: *Sustainable Engineering*, pp. 297–316, 2019, doi: 10.1017/9781108333726.016.
2. Barbanera, M., Castellini, M., Tasselli, G., Turchetti, B., Cotana, F., Buzzini, P., Prediction of the Environmental Impacts of Yeast Biodiesel Production from Cardoon Stalks at Industrial Scale. *Fuel*, 283, May 2020, 118967, 2021, doi: 10.1016/j.fuel.2020.118967.
3. Black, M.J., Borrion, A.L., Mwabonje, O., Chapter 1. Theories and Tools for the Assessment of Environmental Impacts of Human Activities, in: *Life Cycle Assessment*, pp. 1–26, 2021, doi: 10.1039/9781788016209-00001.
4. Cabrera-Jiménez, R., Mateo-Sanz, J.M., Gavaldà, J., Jiménez, L., Pozo, C., Comparing Biofuels through the Lens of Sustainability: A Data Envelopment Analysis Approach. *Appl. Energy*, 307, 2022, doi: 10.1016/j.apenergy.2021.118201.
5. Chopra, J., Rangarajan, V., Rathnasamy, S., Dey, P., Life Cycle Assessment as a Key Decision Tool for Emerging Pretreatment Technologies of Biomass-to-Biofuel: Unveiling Challenges, Advances, and Future Potential. *Bioenergy Res.*, 0123456789, 2024, doi: 10.1007/s12155-024-10741-8.
6. Chopra, J., Rangarajan, V., Sen, R., Recent Developments in Oleaginous Yeast Feedstock Based Biorefinery for Production and Life Cycle Assessment

of Biofuels and Value-Added Products. *Sustain. Energy Technol. Assess.*, 53, PB, 102621, 2022, doi: 10.1016/j.seta.2022.102621.
7. Chopra, J., Tiwari, B.R., Dubey, B.K., Sen, R., Environmental Impact Analysis of Oleaginous Yeast Based Biodiesel and Bio-Crude Production by Life Cycle Assessment. *J. Clean. Prod.*, 271, 122349, 2020, doi: 10.1016/j.jclepro.2020.122349.
8. Codispoti, L.A., The Limits to Growth. *Nature*, 387, 6630, 237–38, 1997, doi: 10.1038/387237a0.
9. Cordero-Lanzac, T., Ramirez, A., Navajas, A., Gevers, L., Brunialti, S., Gandía, L.M., Aguayo, A.T., Mani Sarathy, S., Gascon, J., A Techno-Economic and Life Cycle Assessment for the Production of Green Methanol from CO2: Catalyst and Process Bottlenecks. *J. Energy Chem.*, 68, 255–66, 2022, doi: 10.1016/j.jechem.2021.09.045.
10. Corona, A., Ambye-Jensen, M., Vega, G.C., Hauschild, M.Z., Birkved, M., Techno-Environmental Assessment of the Green Biorefinery Concept: Combining Process Simulation and Life Cycle Assessment at an Early Design Stage. *Sci. Total Environ.*, 635, 100–111, 2018, doi: 10.1016/j.scitotenv.2018.03.357.
11. Finnveden, G. and Potting, J., Life Cycle Assessment, in: *Encyclopedia of Toxicology*, Third Edition, P. Wexler (Ed.), pp. 74–77, Academic Press, Oxford, 2014.
12. Friedlingstein, P., Houghton, R.A., Marland, G., Hackler, J., Boden, T.A., Conway, T.J., Canadell, J.G., Raupach, M.R., Ciais, P., Le Quéré, C., Update on CO2 Emissions. *Nat. Geosci.*, 3, 12, 811–12, 2010, doi: 10.1038/ngeo1022.
13. Gil, A., Challenges on Waste-to-Energy for the Valorization of Industrial Wastes: Electricity, Heat and Cold, Bioliquids and Biofuels. *Environ. Nanotechnol. Monit. Manag.*, 17, 2022, doi: 10.1016/j.enmm.2021.100615.
14. Goedkoop, M., Heijungs, R., Huijbregts, M., De Schryver, A., Struijs, J., Van Zelm, R., ReCiPe 2008. *Potentials*, pp. 1–44, 2009.
15. Johnson, E., Handbook on Life Cycle Assessment Operational Guide to the ISO Standards. *Environ. Impact Assess. Rev.*, 23, 1, 129–30, 2003, doi: 10.1016/s0195- 9255(02)00101-4.
16. Kowalski, Z., Kulczycka, J., Verhé, R., Desender, L., De Clercq, G., Makara, A., Generowicz, N., Harazin, P., Second-Generation Biofuel Production from the Organic Fraction of Municipal Solid Waste. *Front. Energy Res.*, 1–15, 10 August, 2022, doi: 10.3389/fenrg.2022.919415.
17. Kulikova, Y., Ilinykh, G., Sliusar, N., Babich, O., Bassyouni, M., Life Cycle Assessments of Biofuel Production from Beach-Cast Seaweed by Pyrolysis and Hydrothermal Liquefaction. *Energy Convers. Manage.: X*, 23, July, 2024, doi: 10.1016/j.ecmx.2024.100647.
18. Makepa, D.C., Chihobo, C.H., Manhongo, T.T., Musademba, D., Life-Cycle Assessment of Microwave-Assisted Pyrolysis of Pine Sawdust as an Emerging Technology for Biodiesel Production. *Results Eng.*, 101480, 20 July, 2023, doi: 10.1016/j.rineng.2023.101480.

19. Malone, T., Life Cycle Inventory and Analysis of Disposable Hot Beverage Cup Technologies: Coated Paper, Polystyrene Foam, and Expanded Recycled PET. *Society of Plastics Engineers - Global Plastics Environmental Conference 2009, GPEC 2009*, vol. 2, pp. 1025–57, 2009.
20. Osman, A.I., Farghali, M., Rashwan, A.K., Life Cycle Assessment of Biochar as a Green Sorbent for Soil Remediation. *Curr. Opin. Green Sustainable Chem.*, 46, 100882, 2024, doi: 10.1016/j.cogsc.2024.100882.
21. Walter K. and Birgit G., Life Cycle Assessment (LCA): A Guide to Best Practice, First Edition. Wiley-VCH Verlag GmbH & Co. KGaA. Published by Wiley-VCH Verlag GmbH & Co. KGaA, 2014.
22. Srikumar, K., Tan, Y.H., Kansedo, J., Tan, I.S., Mubarak, N.M., Ibrahim, M.L., Yek, P.N.Y., Foo, H.C.Y., Karri, R.R., Khalid, M., A Review on the Environmental Life Cycle Assessment of Biodiesel Production: Selection of Catalyst and Oil Feedstock. *Biomass Bioenergy*, 185, January, 107239, 2024, doi: 10.1016/j.biombioe.2024.107239.
23. Tillman, A.M., Ekvall, T., Baumann, H., Rydberg, T., Choice of system boundaries in life cycle assessment. *J. Clean. Prod.*, 2, 21–29, 1994, ISSN 0959-6526, https://doi.org/10.1016/0959-6526(94)90021-3.
24. Togarcheti, S.C., Mediboyina, M.K., Chauhan, V.S., Mukherji, S., Ravi, S., Mudliar, S.N., Life Cycle Assessment of Microalgae Based Biodiesel Production to Evaluate the Impact of Biomass Productivity and Energy Source. *Resour. Conserv. Recycl.*, 122, 286–94, 2017, doi: 10.1016/j.resconrec.2017.01.008.
25. Yang, Y., A Unified Framework of Life Cycle Assessment. *Int. J. Life Cycle Assess.*, 24, 4, 620–26, 2019, doi: 10.1007/s11367-019-01595-w.

10

Techno-Economic Analysis Perspective of Green Engineering

Harishbabu Balaraman, Karishma Chandrasekaran and Senthilkumar Rathnasamy*

Green Separation Engineering Laboratory, School of Chemical and Biotechnology, SASTRA Deemed to be University, Thanjavur, India

Abstract

Over the last few chapters, various approaches and tools employed in achieving sustainability have been discussed in detail. Although these techniques have their environmental advantages, their adaptation to industrial scale requires weighing the economic challenges in their transition. Thus, the technical and economic feasibility of the alternative green methodology needs to be evaluated. This chapter encapsulates the techno-economic perspective of certain commercially significant products obtained through the sustainable approach. To begin with, a detailed perspective on the economic feasibility of modernized sustainable plant and algal biorefineries is presented in detail. Further, the economic aspects of employing process intensification techniques are also briefly discussed. Finally, the importance of techno-economic analysis on the extraction of commercially important bioactives is also listed. On the whole, this chapter has elaborated on how the economic aspects of a process are influenced by altering the former to achieve sustainability.

Keywords: Techno economic analysis, sustainability, economic feasibility, biorefinary, biofuel, bioactive extraction, energy consumption

10.1 Introduction

Techno-economic assessment (TEA) is a critical step in evaluating the possibility of scaling up a new technology or product from the research scale

---

**Corresponding author*: senthilrathna@sastra.ac.in

Senthilkumar Rathnasamy, Vivek Rangarajan and Umile Gianfranco Spizzirri (eds.) Green Engineering and Sustainable Technology: Tools and Techniques, (265–284) © 2026 Scrivener Publishing LLC

to the industrial scale. This combines the aspects of engineering design, process modeling, and economic feasibility to determine the probability of scaling up a process, system, or its configuration to achieve identical efficiency as achieved in research scale [1]. In other words, TEA is a tool that aids in examining the cost, profits, risks, and uncertainties in scaling up a new or modified technology. In general, a TEA analysis has a wider thought of economic feasibility than technological adaptability. Beyond defining the potential economic bottlenecks of the proposed process, it also suggests ways to alter investment factors by reiterating the choice of process configurations, thereby reducing capital expenditures [2]. To administer a TEA for a process, a sequence of activities is needed to be performed as follows. Initially, the scope of the project needs to be defined, which carries essential information like establishing key performance metrics (yield or purity); calculating operational and capital costs, potential revenue streams and market upstream process (USP), and sensitivity analysis; and assessing the market stability for the product. Following this, a virtual process model with the capacity, influx flow rate, production rate, and product recovery rate was designed. Finally, the evaluation of outcome parameters like net present value (after waste disposal and transport) and rate of return completes the TEA process [3]. Although TEA can be performed for any operation, the core values of sustainable reformation are observed to have significance in two commercially important products (biorefinery and nutraceuticals), thus their exclusion in the current chapter.

TEA of the above-mentioned two processes is greatly influenced by the type of solvents used and the intensification method being adopted. In general, green solvents, such as ionic liquids, supercritical fluids, and bio-based solvents, offer significant advantages over traditional organic solvents. These solvents are characterized by their low toxicity, biodegradability, and high efficiency in biomass processing [4]. By employing these solvents, product yield rises to a greater good, while simultaneously reducing energy consumption and waste generation. On other hand, green extraction techniques provide numerous advantages over traditional methods, particularly in terms of environmental sustainability and efficiency. One of the primary benefits is improved extraction efficiency, yielding higher amounts of product in a shorter time frame. This efficiency is critical for industries that require quick and effective extraction processes. By adopting these innovative techniques, industries can enhance their sustainability efforts while maintaining high-quality product standards.

10.2 Economic Aspects of Biorefinery

The economic feasibility of biorefineries utilizing green technologies depends on several key factors. Initially, substantial capital investments are required for infrastructure development and technology implementation. Ongoing operational costs, including feedstock procurement, processing, labor, and maintenance, also contribute to the overall expense. The fluctuating market prices of biofuels in comparison to fossil fuels further influence the economic viability. Additionally, the integration of green solvents can lead to improved energy efficiency, as they can facilitate reactions at lower temperatures and pressures, thereby reducing energy consumption. However, studies suggest that, although the initial capital costs may be significant, the long-term operational savings and potential market advantages can ultimately lead to favorable economic outcomes.

10.2.1 Economic Assessment of First-Generation Biorefinery

Agro-industrial waste, a by-product of food processing and agricultural activities, presents a significant opportunity for sustainable biofuel production. This waste stream, comprising materials like corn stover, sugarcane bagasse, and fruit peels, is often underutilized and can be transformed into valuable energy resources. By harnessing the potential of these waste materials, we can reduce our reliance on fossil fuels and mitigate climate change. Figure 10.1 shows that the generation development of biorefinery.

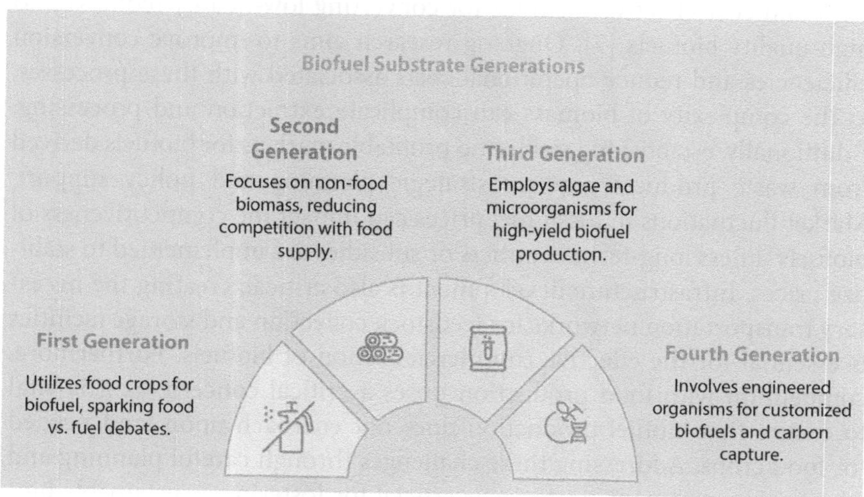

Figure 10.1 Generation development of biorefinery.

This chapter provides a comprehensive techno-economic analysis (TEA) of biofuels derived from agro-industrial waste, exploring conversion technologies, economic benefits, challenges, and the integration of green solvents [5].

The utilization of agro-industrial wastes for biofuel production presents several economic advantages, primarily through reduced feedstock costs. Many of these wastes are by-products from other industries and can often be procured at minimal or no cost, significantly lowering the overall expenses associated with biofuel production compared to conventional feedstocks. Additionally, this approach allows for the diversification of revenue streams; agricultural and industrial sectors that generate these wastes can benefit financially by transforming what would typically be discarded into valuable energy products [6]. This not only provides a new source of income for farmers and processors but also fosters job creation in rural and agricultural communities where these wastes are produced, contributing to local economic development and enhancing community resilience. Furthermore, the production of biofuels from agro-industrial waste helps improve energy security by reducing reliance on imported fossil fuels, enabling nations to diversify their energy sources and promote sustainable practices in the energy sector.

The challenges and prospects of first-generation biorefinery are listed as below. Despite the economic advantages, several challenges hinder the widespread adoption of biofuels from agro-industrial waste. The production of biofuels from agro-industrial waste faces several significant challenges that must be addressed to ensure successful commercialization. One of the primary technological barriers is the development of efficient and cost-effective methods for converting low-quality biomass into high-quality biofuels [7]. Ongoing research aims to improve conversion efficiencies and reduce operational costs associated with these processes, as the complexity of biomass can complicate extraction and processing. Additionally, establishing stable and profitable markets for biofuels derived from waste products requires strategic planning and policy support. Market fluctuations in fossil fuel prices can impact the competitiveness of biofuels unless long-term contracts or subsidies are implemented to stabilize prices. Infrastructure development is also critical; creating the necessary transportation networks for feedstock collection and storage facilities is essential for the effective commercialization of biofuels. Furthermore, competition with food production poses a critical concern, as it is vital to ensure that biofuel production does not encroach upon land needed for food crops. Addressing these challenges through careful planning and resource management strategies is crucial for fostering a sustainable biofuel industry [8].

There are several critical opportunities for innovation in technology development, market expansion, and policy frameworks that support sustainable practices in the utilization of agro-industrial waste for biofuel production. Initially, the use of agro-industrial wastes for biofuel production has reduced landfill and combustion rates, bringing down the global carbon emission levels. On the other hand, the use of green solvents has positively influenced biofuel yield lowering the minimum biodiesel selling price (MBSP). Furthermore, green solvents typically operate at lower temperatures and pressures than conventional methods, which contributes to significant energy savings during processing. These reductions in energy consumption not only decrease operational costs but also align with sustainability goals, making the production process more economically viable [9].

In conclusion, the valorization of agro-industrial waste for biofuel production represents a sustainable approach to addressing energy needs while managing waste effectively. The diverse types of agro-industrial wastes available provide ample opportunities for conversion into renewable energy through various advanced technologies such as gasification, anaerobic digestion, pyrolysis, and fermentation. The economic benefits associated with utilizing these wastes, such as reduced feedstock costs, job creation, diversification of revenue streams, and enhanced energy security, further underscore the viability of this approach within a circular economy framework. Although challenges remain in terms of technological advancements, market development, infrastructure requirements, and competition with food production systems, innovative solutions that leverage green solvents could play a pivotal role in overcoming these barriers. By harnessing agro-industrial wastes effectively through sustainable practices and cutting-edge technologies, we can make significant strides toward achieving a more resilient energy future that minimizes environmental impacts while maximizing resource recovery from agricultural activities.

10.2.2 Economic Assessment of Second Generation Biorefinery

Among various feedstocks, oilseeds have gained significant prominence due to their high oil content and established cultivation practices. This section delves into a comprehensive TEA of biodiesel production from oilseeds, exploring the cost dynamics, market demand, and overall feasibility of this emerging biofuel technology [10]. Oilseeds, including soybean, canola, rapeseed, sunflower, and palm, are cultivated worldwide for their oil-rich seeds, which typically contain between 30% and 50% oil by weight. These seeds serve as a significant feedstock for biodiesel production. The oil extracted from these seeds undergoes a chemical process known as transesterification, wherein triglycerides are converted into fatty acid methyl esters, the primary components of biodiesel.

The TEA of oilseed feedstocks for biodiesel production reveals several advantages that enhance the economic viability of this renewable energy source. This demand is driven by government mandates, subsidies, and an increasing consumer awareness of environmental sustainability. The robust market for biodiesel provides a strong incentive for investments in oilseed cultivation and processing facilities, ensuring a viable market for the produced biofuel. Additionally, the processing of oilseeds generates valuable by-products such as oilseed meal and glycerol. These by-products can be repurposed for various applications, including animal feed and industrial uses, further enhancing the economic viability of biodiesel production by creating additional revenue streams [11].

However, despite these advantages, there are notable disadvantages associated with using oilseed feedstocks that must be considered in a TEA. One significant concern is the volatility of raw material costs. The prices of oilseeds can fluctuate dramatically because of factors such as weather conditions, global market dynamics, and competition with food production. This price volatility poses risks to the economic stability of biodiesel production, as fluctuations in raw material costs directly impact overall production expenses and profitability. Consequently, producers must implement strategies to mitigate these risks, such as securing long-term contracts or diversifying feedstock sources [12].

Finally, processing costs associated with extracting oil from seeds present another challenge in the economic analysis of biodiesel production. The extraction process involves multiple steps—such as seed cleaning, drying, crushing, and extraction—each requiring significant energy inputs and specialized equipment. These processing costs contribute to the overall expenses involved in biodiesel production and must be carefully managed to maintain profitability. Innovations in processing technologies that enhance efficiency or reduce energy consumption could play a pivotal role in improving the economic feasibility of using oilseeds as feedstocks for biodiesel.

In conclusion, a comprehensive TEA of oilseed feedstocks reveals both advantages and disadvantages that influence their viability for biodiesel production. Although high oil content, established cultivation practices, and growing market demand present significant opportunities for profitability, challenges such as raw material cost volatility, land use competition, environmental concerns, and processing costs must be addressed strategically. Although biofuels derived from oilseeds can reduce greenhouse gas emissions compared to fossil fuels, it is important to assess their life cycle emissions, including those from cultivation, processing, and transportation. Furthermore, the expansion of oilseed cultivation can lead to indirect land use change in other regions, as the need for additional agricultural land may drive the conversion of forests or grasslands into farmland to produce other commodities [10]. Addressing these challenges requires a comprehensive approach that balances energy production with environmental sustainability. By understanding these dynamics, stakeholders can make informed decisions that optimize biodiesel production processes while promoting sustainable practices within the industry [13].

Case Study Insights

Several studies have conducted techno-economic analyses of biodiesel production from various oilseed feedstocks.

- **Soybean Biodiesel:** Studies have indicated that the cost of producing biodiesel from soybean oil can vary significantly depending on factors such as feedstock prices, production scale, and regional variations in input costs. However, typical production costs for soybean biodiesel range from $0.90 to $1.20 per liter.
- **Canola Biodiesel:** Due to its higher oil content and often lower production costs, canola has been shown to offer a more favorable economic profile than soybean biodiesel. Studies have estimated the production cost of canola biodiesel to range from $0.80 to $1.00 per liter.
- **Palm Oil Biodiesel:** Palm oil, with its high oil yield and relatively low production costs, has emerged as a highly cost-effective feedstock for biodiesel production. Studies have indicated that palm oil biodiesel can be produced at a cost of around $0.70 to $0.80 per liter, making it one of the most economically attractive options.

10.2.3 Economic Assessment of Third-Generation Biorefinery

Algal biomass has emerged as a promising feedstock for biofuel production due to its rapid growth rate, high lipid content, and potential to utilize non-arable land and wastewater. This versatile resource offers a sustainable alternative to traditional fossil fuels. As the world seeks to transition to cleaner energy sources, algal biofuels present a viable solution that aligns with environmental sustainability and energy security goals. Algae can utilize nutrients from wastewater during cultivation, reducing pollution while providing a valuable resource for biofuel production. This approach aligns with sustainable practices and circular economy principles by transforming waste into energy. The TEA of algal biomass for biodiesel production reveals significant insights when comparing biomass sourced from natural resources versus cultivated algae [14]. Algal biomass, particularly from species like *Chlorella* and *Spirulina*, offers a promising feedstock due to its high lipid content and rapid growth rates. Cultivating algae in controlled environments, such as photobioreactors, allows for optimized growth conditions that can yield substantial quantities of oil suitable for biodiesel production. This method contrasts sharply with the collection of biomass from natural resources, which may be subject to seasonal variations and ecological constraints. The cultivation approach not only enhances predictability in yield but also minimizes competition with food crops, as algae can thrive in non-arable land and utilize saline or brackish water.

In terms of economic viability, the cost structure associated with algal biomass cultivation is a critical factor. Recent studies indicate that the production costs for cultivated algal biomass can be significantly higher than those derived from traditional agricultural sources due to the need for specialized infrastructure and controlled growth conditions. For instance, the TEA conducted on *Chlorella* sp. in Indonesia highlighted that the internal rate of return for biodiesel production from cultivated algae is around 3%, which underscores the economic challenges faced by this sector [15]. In contrast, biomass sourced from established agricultural practices tends to have lower raw material costs, although it may not offer the same environmental benefits as algal cultivation. The processing phase also presents distinct differences between cultivated algal biomass and traditional biomass sources. Algal biodiesel production typically involves processes such as transesterification, where the oil extracted from algae is converted into biodiesel using catalysts like sodium hydroxide. This process can yield valuable co-products such as glycerol, which can further enhance the economic feasibility of algal biodiesel production. However, the extraction and processing of algal oil require significant energy inputs and advanced technology, potentially increasing operational costs compared to processing terrestrial biomass, which may involve simpler extraction methods.

Another critical aspect of TEA is the potential for co-product valorization in algal biodiesel production. The ability to extract and utilize by-products like proteins and carbohydrates from algal biomass can significantly offset production costs and enhance overall profitability. For example, integrating a combined algal processing approach allows for the fractionation of biomass into lipid-rich oils and carbohydrate-rich fractions that can be converted into bioethanol or other biochemicals. This multi-product strategy contrasts with traditional biomass processing methods that often focus solely on oil extraction, limiting revenue streams and increasing reliance on single-product sales.

Despite the advantages of algal biomass cultivation, challenges remain regarding scalability and cost-effectiveness. The initial capital investment required for setting up algae cultivation systems can be substantial, and ongoing operational costs must be carefully managed to ensure economic viability. Additionally, achieving economies of scale is essential for reducing the minimum fuel selling price of algal biodiesel to compete with petroleum-based fuels effectively [16]. As research continues to optimize cultivation techniques and processing methods, addressing these challenges will be crucial for advancing the commercialization of algal biofuels and realizing their potential as a sustainable energy source. As research continues to advance our understanding of algal biology and processing technologies, the potential for algal biomass as a sustainable solution for meeting global energy demands becomes increasingly promising. By leveraging these advancements alongside supportive policies aimed at fostering renewable energy development, we can unlock the full potential of algal biofuels in contributing toward sustainable energy solutions that benefit both society and the environment.

10.2.4 Role of Green Solvents in Economic Perspective of Biorefinery

The integration of green solvents in the extraction of lipids from algal biomass for biodiesel production offers a promising approach to enhance both sustainability and economic viability. Traditional lipid extraction methods often rely on volatile organic compounds, such as hexane, which pose environmental risks and require significant energy for recovery. In contrast, the use of bio-based solvents, including deep eutectic solvents (DESs) and ionic liquids, has demonstrated improved lipid extraction efficiency while reducing the environmental footprint [17]. Research indicates that solvents like ethyl lactate and 2-methyltetrahydrofuran can achieve lipid extraction rates that are two to three times higher than those obtained with hexane. This increased efficiency not only enhances overall biodiesel yield but also improves the quality of the extracted lipids, making them more suitable for biodiesel production [18].

From a techno-economic perspective, the adoption of green solvents can significantly impact production costs and market demand for algal biodiesel. Although the initial investment in green solvent technology may be higher compared to traditional methods, the long-term benefits include reduced processing costs associated with lower energy consumption and improved extraction efficiency. Enhanced lipid yields contribute to a more favorable MBSP, making algal biodiesel more competitive in the market. As consumers and regulatory bodies increasingly prioritize sustainability, the demand for biodiesel produced using eco-friendly methods is expected to rise. This shift in market dynamics could lead to greater acceptance and preference for biodiesel derived from algal sources processed with green solvents. The incorporation of green solvents into biorefineries presents a promising pathway toward sustainable and economically viable biofuel production [19]. Although challenges remain, such as high initial costs and fluctuating market dynamics, the long-term benefits in terms of environmental sustainability and economic competitiveness make this approach a compelling option for the future of energy production.

Moreover, the use of green solvents facilitates the valorization of co-products generated during lipid extraction. By selectively extracting valuable compounds such as proteins and carbohydrates alongside lipids, producers can create additional revenue streams that enhance the overall economic viability of algal biodiesel production. This ability to utilize co-products not only improves profitability but also aligns with circular economy principles by minimizing waste and maximizing resource use. Although challenges remain regarding scalability and cost-effectiveness in industrial applications, continued investment in research and development will be essential to refine these technologies. Ultimately, integrating green solvents into algal biomass processing can transform waste materials into valuable products and contribute to a more sustainable biofuel industry that meets contemporary environmental goals and market demands.

10.3 Economic Aspects of Bioactive Extraction

The world is increasingly turning to nature for solutions to health and wellness. From ancient herbal remedies to modern pharmaceuticals, natural products have always played a crucial role. To extract the valuable bioactive compounds from these natural sources, scientists and researchers have traditionally relied on organic solvents. However, these solvents can be harmful to the environment and human health. A more sustainable and eco-friendly approach is emerging: the use of green solvents [20]. These solvents are less toxic, biodegradable, and often derived from renewable resources. By employing green solvents, we can extract bioactive compounds more efficiently and safely. Although green extraction techniques offer numerous benefits, a comprehensive TEA is crucial to assess their feasibility. This analysis considers factors such as capital investment, operational costs, energy consumption, and product yield. By carefully evaluating these factors, researchers and industries can make informed decisions about the most suitable green extraction methods for specific applications. As we continue to explore the potential of natural products, green extraction techniques will play a pivotal role in ensuring a sustainable future. By adopting these eco-friendly approaches, we can harness the power of nature to improve human health and protect our planet [21].

10.3.1 Economic Perspective in Microwave-Based Extraction

Microwave extraction (ME) has emerged as a powerful technique for extracting bioactive compounds from plant materials. This innovative method leverages microwave radiation to enhance the efficiency of extraction processes, offering several advantages over traditional techniques, including increased efficiency, reduced processing time, and lower solvent usage. This chapter provides a detailed examination of ME, exploring its mechanisms, benefits, economic considerations, real-world applications, and its role in sustainable practices.

Microwave-assisted extraction (MAE) has emerged as a revolutionary technique for extracting bioactive compounds from various biomass sources, including agro-industrial waste. This method leverages microwave energy to heat solvents rapidly, which enhances the solubilization of valuable compounds from plant matrices. From a techno-economic perspective, MAE offers several advantages over traditional extraction methods, such as reduced extraction time and lower solvent consumption. The efficiency of MAE can lead to higher yields of bioactive compounds, which is crucial for meeting market demands for functional ingredients in food, cosmetics, and pharmaceuticals. By optimizing extraction parameters—such as microwave power, extraction time, and solvent concentration—producers can maximize the yield and quality of the extracted compounds while minimizing costs associated with raw materials and energy. The importance of MAE in the market is underscored by its alignment with sustainability trends and consumer preferences for environmentally friendly products [22]. Traditional extraction methods often rely on harsh solvents that pose environmental risks and require extensive safety measures during handling and disposal. In contrast, MAE can utilize green solvents that are less toxic and more biodegradable, reducing the overall ecological footprint of the extraction process. This shift not only meets regulatory demands for lower emissions but also enhances the marketability of bioactive extracts derived from agro-industrial waste. As consumers become increasingly aware of the environmental impacts of their purchases, products that are extracted using sustainable methods are likely to gain a competitive edge in the marketplace [23].

Moreover, the economic implications of MAE extend beyond extraction efficiency to include the recovery and valorization of by-products generated during the extraction process. The ability to extract multiple bioactive compounds simultaneously allows producers to diversify their product offerings and create additional revenue streams. For instance, alongside primary bioactive compounds, co-products such as phenolics or flavonoids can be extracted and marketed for their antioxidant properties. This multi-product approach not only enhances profitability but also aligns with circular economy principles by maximizing resource utilization and minimizing waste. By effectively utilizing all components of agro-industrial waste, businesses can improve their overall economic viability while contributing to sustainability goals.

Despite its advantages, challenges remain regarding the scalability and cost-effectiveness of implementing MAE in industrial applications. Although laboratory-scale studies demonstrate promising results, transitioning to commercial-scale operations requires careful consideration of factors such as solvent recovery and process optimization. Additionally, achieving consistent results across varying biomass types may necessitate further research into the specific parameters that influence extraction efficiency. Continued investment in technology development will be essential for refining MAE processes and ensuring their economic feasibility in large-scale production settings. By addressing these challenges, stakeholders can fully realize the benefits of MAE as a viable method for producing high-quality bioactive compounds from agro-industrial waste while meeting market demands for sustainable products. In summary, MAE offers numerous techno-economic advantages over traditional methods, including reduced extraction time, lower solvent consumption, improved product quality, and enhanced economic viability through by-product recovery. As industries continue to seek more efficient and sustainable extraction techniques, MAE stands out as a promising solution that meets contemporary market demands while addressing environmental concerns.

10.3.2 Economic Aspects of Ultrasound-Assisted Extraction

Ultrasound-assisted extraction (UAE) is gaining recognition as an innovative technique for extracting bioactive compounds from diverse biomass sources, including agro-industrial waste. Similar to MAE, UAE utilizes energy to enhance the extraction process, but it does so through ultrasonic waves, which create cavitation bubbles in the solvent. This phenomenon leads to increased mass transfer and improved solubilization of bioactive compounds from plant matrices. The integration of green solvents in UAE processes represents a notable advancement in the extraction of bioactive compounds. This innovative approach utilizes solvents such as water, ethanol, and DESs to significantly enhance extraction yields. Research has demonstrated that combining UAE with supercritical fluid extraction (SFE) using these green solvents can produce higher concentrations of bioactive compounds, including phenolics, compared to traditional organic solvent methods [24]. The unique ability of green solvents to dissolve a wide variety of bioactive compounds is essential for optimizing extraction processes, while the cavitation effect generated by ultrasonic waves effectively disrupts plant cell walls, promoting better solvent penetration and improved contact with the desired compounds.

A key advantage of employing green solvents in UAE is their reduced environmental impact. Conventional organic solvents are often associated with toxicity risks and pollution during extraction processes. In contrast, green solvents adhere to the principles of green chemistry, which emphasize minimizing hazardous substances in chemical processes. Transitioning to environmentally friendly extraction methods not only mitigates waste but also enhances the sustainability profile of industries such as food and pharmaceuticals. This shift aligns with regulatory standards and growing consumer demand for sustainable products. Moreover, green extraction methods typically require less energy due to shorter extraction times and lower operational temperatures [25]. The combination of UAE with green solvents disrupts plant cell structures efficiently, leading to faster release and solubilization of bioactive compounds. This reduction in energy consumption not only lowers operational costs but also contributes to a smaller carbon footprint for the extraction process. Techno-economic analyses indicate that although initial investments in green technologies may be higher, long-term savings from decreased energy use and solvent recovery can render these methods economically viable.

The milder conditions associated with green solvent extraction are crucial for maintaining the integrity and bioactivity of sensitive compounds. For instance, lower temperatures during extraction help prevent the thermal degradation of heat-sensitive bioactive substances. This preservation is vital for industries that depend on high-quality extracts for their health benefits. By maintaining bioactivity, producers can enhance product value and foster consumer trust in natural products. Economic assessments suggest that products derived from preserved bioactive compounds can command higher market prices, benefiting producers in the long run. The choice of solvent is pivotal in determining the efficiency of UAE processes [26]. Different bioactive compounds exhibit varying solubility profiles based on their chemical nature; thus, selecting appropriate green solvents tailored to specific classes of compounds is essential. The techno-economic implications indicate that optimizing solvent combinations can lead to improved yields and reduced costs associated with raw material processing. Ongoing research into solvent properties and their interactions with plant matrices remains an active area of investigation, promising further enhancements in extraction efficiency.

Looking ahead, the application of green solvents in UAE processes offers a promising pathway for advancing sustainable practices in the field of bioactive compound extraction. Continued research focused on optimizing solvent combinations and refining extraction parameters will enhance both the effectiveness and applicability of these technologies. As industries increasingly strive for sustainability, incorporating techno-economic analyses into research will provide valuable insights into the cost-effectiveness and environmental benefits associated with adopting green extraction methods [27]. This comprehensive approach not only supports scientific advancement but also aligns with global trends toward sustainability in food production and pharmaceuticals. By leveraging the advantages of green solvents within UAE frameworks, stakeholders can contribute significantly to environmental conservation while meeting consumer demands for high-quality, sustainable products.

10.3.3 Economic Aspects of Supercritical Fluid Extraction

SFE has gained prominence as an innovative method for extracting bioactive compounds from natural products, primarily utilizing supercritical carbon dioxide (CO_2) as the solvent. This technique leverages the unique properties of CO_2 in its supercritical state, allowing it to act as both a gas and a liquid, which enhances its ability to solvate both non-polar and polar compounds. The process of SFE involves adjusting temperature and pressure conditions to bring CO_2 into a supercritical state [28]. In this state, CO_2 exhibits low viscosity and high diffusivity, enabling it to penetrate solid matrices effectively while selectively dissolving target bioactive compounds. This selectivity is a key advantage of SFE, as it allows for the extraction of specific compounds based on their solubility characteristics. This chapter delves into the mechanisms of SFE, its advantages, economic considerations, and real-world applications, providing a comprehensive TEA based on recent research findings.

Advantages of Supercritical Extraction

Selectivity

One of the most significant benefits of SFE is its ability to selectively extract specific bioactive compounds by manipulating temperature and pressure parameters. For instance, varying these conditions can enhance the solubility of certain phytochemicals while minimizing the extraction of undesired components [29]. This capability is particularly valuable in industries where purity and specificity are paramount, such as pharmaceuticals and nutraceuticals.

No Residual Solvents

Another critical advantage of SFE is that CO_2 is non-toxic and leaves no harmful residues in the extracted products. This characteristic aligns well with food safety regulations and consumer preferences for clean-label products. The absence of residual solvents not only enhances product safety but also simplifies the processing steps required for product purification.

Environmental Benefits

The use of CO_2 as a solvent significantly reduces environmental impacts compared to traditional organic solvents. Conventional extraction methods often rely on solvents that can be hazardous to human health and the environment. In contrast, SFE minimizes these risks by utilizing a naturally occurring gas that can be easily removed from extracted products through depressurization, leaving behind pure extracts.

High Initial Investment

One of the primary challenges associated with SFE is the high capital costs associated with acquiring and maintaining specialized equipment. The initial investment for SFE systems can be substantial due to the need for high-pressure vessels, pumps, and monitoring systems. However, these costs can be offset by long-term savings on solvent recovery and waste disposal. Additionally, advancements in technology are gradually reducing equipment costs, making SFE more accessible to smaller producers.

High Yield Potential

Despite the initial investment challenge, SFE has demonstrated high yield potential for various bioactive compounds such as carotenoids, essential oils, and phytosterols. The ability to achieve high yields enhances its economic viability in commercial applications. For example, studies have shown that SFE can extract valuable compounds from diverse sources like rice bran oil and ginseng with superior quality compared to conventional methods [30].

Recycling CO_2

The ability to recycle CO_2 within closed-loop systems further enhances sustainability while reducing operational costs over time. By capturing and reusing CO_2 in extraction processes, companies can minimize their environmental footprint while maximizing resource efficiency. This closed-loop approach not only contributes to cost savings but also aligns with global sustainability goals. We hereby conclude by stating that the exploration of SFE highlights significant advancements in sustainable practices within the field of bioactive compound recovery. The unique properties of supercritical CO_2 enable selective extraction while minimizing environmental impacts associated with traditional organic solvents [31].

As industries increasingly seek sustainable alternatives to conventional extraction methods that rely on hazardous solvents, integrating green technologies like SFE will play a critical role in shaping future practices in bioactive compound production. By leveraging these innovative approaches alongside supportive policies aimed at fostering renewable energy development, we can unlock the full potential of bioactive extractions as part of a sustainable economy. In summary, although SFE presents challenges such as high initial investments, its numerous advantages—including selectivity, lack of residual solvents, environmental benefits, high yield potential, and recycling capabilities—position it as a leading method for extracting valuable bioactive compounds from natural products. Continued research and technological advancements will further enhance its economic viability and sustainability in various applications across industries [32].

10.4 Conclusion

This chapter has elaborated the economic perspective of sustainable approaches for obtaining two commercially significant products, i.e., biofuels and bioactive metabolites. Although sustainability has taken a great toll in reducing waste generation and improving process efficiency, their affordability is greatly influenced by the large capital cost required for the specialized facility required for the process. However, enhancing the facility not only guarantees sustainability but also improves the performance (yield and purity) of the product while reducing the time interval, resulting in the modified latter being inclusively economical. Green solvents are the backbone of achieving sustainability due to their recyclable and non-toxic nature. Large amount of works have established the inert nature of green solvents and hence this chapter outlines only the economic prospects of using these sustainable solvents. Conclusively, both sustainability and affordability have a major role to play in scaling up a process and thus need to be investigated inclusively.

AI disclosure statement: The author confirms that no artificial intelligence (AI)-generated content or AI-assisted writing tools were used in the preparation of this chapter (including summaries and the abstract). All text, analysis, and conclusions are the result of the author's original work.

References

1. Lo, S.L.Y., How, B.S., Leong, W.D., Teng, S.Y., Rhamdhani, M.A., Sunarso, J., Techno-economic analysis for biomass supply chain: A state-of-the-art review. *Renew. Sustain. Energy Rev.*, 135, 110164, 2021.
2. Sekoai, P.T., Chunilall, V., Msele, K., Buthelezi, L., Johakimu, J., Andrew, J., *et al.*, Biowaste biorefineries in South Africa: Current status, opportunities, and research and development needs. *Renew. Sustain. Energy Rev.*, 188, 113870, 2023.
3. Kumar Mishra, R., Jaya Prasanna Kumar, D., Narula, A., Minnat Chistie, S., Ullhas Naik, S., Production and beneficial impact of biochar for environmental application: A review on types of feedstocks, chemical compositions, operating parameters, techno-economic study, and life cycle assessment. *Fuel*, 343, 127968, 2023.
4. Azzouz, A. and Hayyan, M., Techno-economic feasibility analysis: The missing piece in the puzzle of deep eutectic solvents. *Sustain. Mater. Technol.*, 39, e00795, 2024.
5. Preethi, M.G., Kumar, G., Karthikeyan, O.P., Varjani, S., RB, J., Lignocellulosic biomass as an optimistic feedstock for the production of biofuels as valuable energy source: Techno-economic analysis, Environmental Impact Analysis, Breakthrough and Perspectives. *Environ. Technol. Innov.*, 24, 102080, 2021.
6. Gnansounou, E. and Dauriat, A., Techno-economic analysis of lignocellulosic ethanol: A review. *Bioresour. Technol.*, 101, 13, 4980–91, 2010.
7. Patel, M., Zhang, X., Kumar, A., Techno-economic and life cycle assessment on lignocellulosic biomass thermochemical conversion technologies: A review. *Renew. Sustain. Energy Rev.*, 53, 1486–99, 2016.
8. Piccolo, C. and Bezzo, F., A techno-economic comparison between two technologies for bioethanol production from lignocellulose. *Biomass Bioenergy*, 33, 3, 478–91, 2009.

9. Martinez-Hernandez, E., Ramírez-Verduzco, L.F., Amezcua-Allieri, M.A., Aburto, J., Process simulation and techno-economic analysis of bio-jet fuel and green diesel production — Minimum selling prices. *Chem. Eng. Res. Des.*, 146, 60–70, 2019.
10. Yong, K.J. and Wu, T.Y., Second-generation bioenergy from oilseed crop residues: Recent technologies, techno-economic assessments and policies. *Energy Convers. Manage.*, 267, 115869, 2022.
11. Kichonge, B. and Kivevele, T., Viability of non-edible oilseed plants and agricultural wastes as feedstock for biofuels production: a techno-economic review from an African perspective. *Biofuels, Bioprod. Biorefin.*, 17, 5, 1382–410, 2023.
12. Diniz, A.P.M.M., Sargeant, R., Millar, G.J., Stochastic techno-economic analysis of the production of aviation biofuel from oilseeds. *Biotechnol. Biofuels*, 11, 1, 161, 2018.
13. Tibesigwa, T., Olupot, P.W., Kirabira, J.B., The critical techno-economic aspects for production of B10 biodiesel from second generation feedstocks: a review. *Int. J. Sustain. Energy*, 41, 7, 751–71, 2022.
14. Nagarajan, S., Chou, S.K., Cao, S., Wu, C., Zhou, Z., An updated comprehensive techno-economic analysis of algae biodiesel. *Bioresour. Technol.*, 145, 150–6, 2013.
15. Quinn, J.C. and Davis, R., The potentials and challenges of algae based biofuels: A review of the techno-economic, life cycle, and resource assessment modeling. *Bioresour. Technol.*, 184, 444–52, 2015.
16. Maroušek, J., Gavurová, B., Strunecký, O., Maroušková, A., Sekar, M., Marek, V., Techno-economic identification of production factors threatening the competitiveness of algae biodiesel. *Fuel*, 344, 128056, 2023.
17. Kumar, A.K., Sharma, S., Dixit, G., Shah, E., Patel, A., Boczkaj, G., Techno-economic evaluation of a natural deep eutectic solvent-based biorefinery: Exploring different design scenarios. *Biofuels, Bioprod. Biorefin.*, 14, 4, 746–63, 2020.
18. Chovau, S., Degrauwe, D., Van der Bruggen, B., Critical analysis of techno-economic estimates for the production cost of lignocellulosic bio-ethanol. *Renew. Sustain. Energy Rev.*, 26, 307–21, 2013.
19. Soh, L. and Eckelman, M.J., Green Solvents in Biomass Processing. *ACS Sustain. Chem. Eng.*, 4, 11, 5821–37, 2016.
20. Vázquez-Romero, B., Perales, J.A., Pereira, H., Barbosa, M., Ruiz, J., Techno-economic assessment of microalgae production, harvesting and drying for food, feed, cosmetics, and agriculture. *Sci. Total Environ.*, 837, 155742, 2022.
21. Balaraman, H., Selvasembian, R., Rangarajan, V., Rathnasamy, S., Sustainable and Green Engineering Insights on Deep Eutectic Solvents toward the Extraction of Nutraceuticals. *ACS Sustain. Chem. Eng.*, 9, 34, 11290–313, 2021.

22. Barrios, C., Fernández-Delgado, M., López-Linares, J.C., García-Cubero, M.T., Coca, M., Lucas, S., A techno-economic perspective on a microwave extraction process for efficient protein recovery from agri-food wastes. *Ind. Crops Prod.*, 186, 115166, 2022.
23. Zapata-Boada, S., Gonzalez-Miquel, M., Jobson, M., Cuéllar-Franca, R.M., A Methodology to Evaluate Solvent Extraction-Based Processes Considering Techno-Economic and Environmental Sustainability Criteria for Biorefinery Applications. *Ind. Eng. Chem. Res.*, 60, 45, 16394–416, 2021.
24. Siddiqui, S.A., Ali Redha, A., Salauddin, M., Harahap, I.A., Rupasinghe, H.P.V., Factors Affecting the Extraction of (Poly)Phenols from Natural Resources Using Deep Eutectic Solvents Combined with Ultrasound-Assisted Extraction. *Crit. Rev. Anal. Chem.*, 55, 1, 139–60, 2025.
25. Vanessa Caicedo Paz, A., Rigano, F., Cafarella, C., Tropea, A., Mondello, L., Paul Martinez Galan, J., *et al.*, Process intensification of ultrasound assisted deep eutectic solvent-based extraction of astaxanthin-rich extract derived from the non-conventional bacterium Paracoccus carotinifaciens. *Sep. Purif. Technol.*, 339, 126674, 2024.
26. Dhar, B.R., Nakhla, G., Ray, M.B., Techno-economic evaluation of ultrasound and thermal pretreatments for enhanced anaerobic digestion of municipal waste activated sludge. *Waste Manage. (Oxford)*, 32, 3, 542–9, 2012.
27. Sicaire, A.-G., Vian, M.A., Fine, F., Carré, P., Tostain, S., Chemat, F., Ultrasound induced green solvent extraction of oil from oleaginous seeds. *Ultrason. Sonochem.*, 31, 319–29, 2016.
28. Weiland, N.T. and White, C.W., Techno-economic analysis of an integrated gasification direct-fired supercritical CO2 power cycle. *Fuel*, 212, 613–25, 2018.
29. Knez, Z., Markočič, E., Leitgeb, M., Primožič, M., Knez Hrnčič, M., Škerget, M., Industrial applications of supercritical fluids: A review. *Energy*, 77, 235–43, 2014.
30. Maio, D.V.D., Boccitto, A., Caruso, G., Supercritical carbon dioxide applications: features and advantages. *2015 5th International Youth Conference on Energy (IYCE)*, pp. 1–4, 2015.
31. Koytsoumpa, E.I., Bergins, C., Kakaras, E., The CO2 economy: Review of CO2 capture and reuse technologies. *J. Supercrit. Fluids*, 132, 3–16, 2018.
32. Mahmud, R., Moni, S.M., High, K., Carbajales-Dale, M., Integration of techno-economic analysis and life cycle assessment for sustainable process design – A review. *J. Clean. Prod.*, 317, 128247, 2021.

11

Conclusion and Future Trends in Sustainability

Dhandapani Ramesh, Gracia Saral Gladison and Senthilkumar Rathnasamy*

Green Separation Engineering Laboratory, School of Chemical and Biotechnology, SASTRA Deemed to be University, Thanjavur, India

Abstract

Green sustainable engineering is one of the backbones of the Circular Economy through sustainable development through waste minimization, increased efficiency, and clean technologies. Urban city such as Freiburg demonstrate their city planning capacity through renewable energy and environmentally friendly transport, while organizations such as IKEA and Dell demonstrate business-driven circular practices and more environmentally friendly product development. Global cooperation, policy enabling, and learning green chemistry through UNEP programs lead to its implementation. Emerging technologies such as Deep Eutectic Solvents (DESs) and Deep Eutectic Electrolytes (DEEs) offer cheaper, safer methods of CO_2 capture and future storage. Process intensification with the aid of AI, nanotechnology, and novel materials increases efficiency and reduces the environmental footprint at the industrial level. These include some such as uses of green nanoemulsions in the food, cosmetic, and pharmaceutical industries; green methods of glycan separation in biomedicine; and green fermentation technology for bio-based products. Techniques like Life Cycle Assessment (LCA) and techno-economic analysis (TEA) equally assess both environmental and economic sustainability and are scalable and competitive. Ultimately, this book will help to foster collaboration among researchers, policymakers, and business leaders to solve environmental issues and also create a low-carbon, resource-efficient economy that promotes innovation, public health, and global progress toward the Sustainable Development Goals.

**Corresponding author*: senthilrathna@sastra.ac.in

Senthilkumar Rathnasamy, Vivek Rangarajan and Umile Gianfranco Spizzirri (eds.) Green Engineering and Sustainable Technology: Tools and Techniques, (285–290) © 2026 Scrivener Publishing LLC

***Keywords*:** Green sustainable engineering, circular economy, emerging technology, process intensification, life cycle assessment, techno-economic analysis, sustainable development goals

Green engineering plays a crucial role in the circular economy (CE), emphasizing its complexity and noteworthy influence on environmental sustainability in a range of industries [1]. Using successful case studies from cities like Freiburg, which represents as a prime model of how to incorporate green engineering into urban planning with its commitment to ecological development, public transit, and renewable energy [2]. Businesses like IKEA [3] and Dell Technologies [4] provide examples of how companies may embrace sustainable practices through closed-loop processes, long-lasting products, and creative design. In order to advance green engineering, global sustainability goals, interdisciplinary collaboration, cutting-edge technology, and supportive policies are crucial. The continuous development of green engineering is essential for promoting economic growth and social well-being in addition to addressing the environmental issues brought on by contemporary industrial activities [5].

The importance of raising public awareness of green chemistry develops inclusive frameworks and strengthens international collaborations. Governments, academic institutions, and corporations can work together efficiently to distribute resources and exchange knowledge to advance green chemistry initiatives. Stakeholders can traverse sustainable practices using clear frameworks created by the UNEP's Green and Sustainable Chemistry effort [6]. Furthermore, educating the public on the advantages of green chemistry with an emphasis on how it promotes safety, health, and environmental preservation inspires community support [7]. Sustainable development objectives can be further advanced by putting these tactics into practice while taking lessons from the past.

Sustainable energy solutions and environmental challenges are receiving more and more attention from the scientific community. Ionic liquids (ILs) and deep eutectic solvents (DESs) can meet energy demands and also reducing harmful effects on the environment, according to a recent study. Development of DESs to enhance their qualities and broaden their application in industries because of their cost-effectiveness compared to ILs has a significant advantage. The efficiency of DESs in capturing carbon dioxide can replace conventional amines utilized in solvent absorption [8]. Physisorption has an advantage with DESs in reducing the energy inputs and has an energy-saving process. These potential applications of DESs promise to shape the future of sustainable energy practices. Green engineering's process intensification emphasizes how important it is to build a sustainable future. Process intensification is the result of combining environmental care, resource optimization, and efficiency. It is becoming more important in a variety of industries because it improves production efficiency and product quality while reducing waste, energy use, and emissions [9]. The incorporation of cutting-edge technology enhanced materials, machine learning, and artificial intelligence to the future of process intensification. These developments hold the potential to create completely new, profitable, and ecologically friendly procedures. Researchers, business executives, and legislators must work together to accomplish these objectives. While industry professionals should place a higher priority on investing in sustainable technology and practices, researchers must use interdisciplinary relationships to push the boundaries of knowledge. Legislative frameworks also need to change to encourage creativity and make it easier for new ideas to be adopted. Stakeholders may steer green engineering toward a future in which process intensification is essential to sustainable development by adopting these concepts, striking a balance between industrial expansion and environmental preservation for future generations [10].

The green nanoemulsion details the main concept of green principles and sustainable synthesis of nanoemulsion, which creates a breakthrough in sustainable development and offering environmentally friendly products, especially in the food, cosmetic, and pharmaceutical industries. These creative ideas improve product performance, safety, and environmental sustainability by utilizing the concepts of green chemistry. With developments in green chemistry and nanotechnology positioned to spearhead this transition, the future success of green nanoemulsions depends on ongoing innovation in environmentally acceptable components, scalable production techniques with more industrial applications [11]. The cutting-edge eco-conscious technologies for glycan separation investigates advanced environmentally friendly glycan separation technologies, mainly focusing on their significance and potential impact on glycomics research. It addresses various topics, including biocatalytic processes, membrane technologies, chromatographic techniques, and green solvents. Despite the distinct advantages that improve energy efficiency and reduce hazardous waste through green technology processes, there are still concerns about scalability, economic viability, and regulatory compliance. Here, we state that glycan separation techniques can be improved by merging many green technologies, resulting in more accurate study results. Innovation in glycan separation will be fueled by upcoming advancements in nanotechnology and biomimetic techniques as well as greater stakeholder collaboration. A culture of sustainability in glycomics research is essential for achieving greater efficiency and scientific discovery in green engineering concepts and cutting-edge technologies [12].

Sustainable fermentation provides a clear review of sustainable fermentation technology, including its principles and numerous applications in a variety of product categories such as organic acids, enzymes, secondary metabolites, pigments, and aroma compounds. It indicates the comparative advantages of sustainable techniques over conventional production methods based on their environmental and economic benefits. It also highlights the importance of using sustainable fermentation technology to tackle global issues such as resource depletion and environmental degradation. These fermentation methodologies play an efficient role in establishing a sustainable economy and minimizing greenhouse gas emissions by increasing the usage of renewable materials with the help of microorganisms and renewable resources. The future of sustainable fermentation technology seems promising, because of advances in microbial genetics, process optimization, and enhanced separation and purification techniques [13]. These advancements promise not just to increase the efficiency of fermentation-based production but also to line with global sustainability targets. As these technologies improve, they have the potential to revolutionize industrial processes, resulting in more sustainable practices and a good environmental impact. The deep eutectic electrolytes (DEEs) detailed the significant advances in developing more sustainable and efficient energy storage systems. These electrolytes, which are made up of environmentally safe and readily available compounds, have exceptional qualities such as ionic conductivity, thermal stability, and tunability. This makes them ideal for the next-generation batteries and supercapacitors. It also suggests some DEEs improve performance parameters such as capacitance, energy density, and cycle stability. Furthermore, DEEs offer a compelling environmentally acceptable alternative to typical electrolytes, such as those found in aqueous systems and ILs derived from renewable sources. The DEEs are valuable contributors to the evolution of sustainable energy technologies [14]. In the future, enormous amounts of R&D will be expected to unlock even more potential in enhancing the efficiency and longevity of energy storage devices.

The life cycle assessment (LCA) for environmental impact analysis states the importance of LCA and its crucial role in analyzing the entire life cycle of the product. It has evolved as a critical systemic tool in a variety of technological development sectors, including medicines, waste to energy, automotive, food production, and biofuels. LCA explores the whole life cycle of a product by evaluating and analyzing the consequences of all procedures within a system that also takes into account local constraints and special conditions. This comprehensive technique not only outlines the issues connected with each stage of the LCA analysis but also makes suggestions for improvement. Furthermore, a techno-economic analysis can supplement LCA by providing information on the economic viability of the analyzed processes. The techno-economic analysis chapter describes the techno-economic perspective of certain commercially significant products obtained through the sustainable approach and perspective on the economic feasibility of modernized sustainable plant and algal biorefineries [15]. Further, the economic aspects of employing process intensification techniques are also briefly discussed. The importance of techno-economic analysis on the extraction of commercially important bioactives is also listed. On the whole, this chapter detailed how the economic aspects of a process get influenced by altering the former to achieve sustainability. Through these chapters, we suggested the implementation of sound green technologies in various fields and how it works to attain sustainable environmental development in the current world [16].

AI disclosure statement: The author confirms that no artificial intelligence (AI)-generated content or AI-assisted writing tools were used in the preparation of this chapter (including summaries and the abstract). All text, analysis, and conclusions are the result of the author's original work.

References

1. D'amato, D. and Korhonen, J., Integrating the green economy, circular economy and bioeconomy in a strategic sustainability framework. *Ecol. Econ.*, 188, 107143, 2021.
2. ICLEI, Freiburg im Breisgau, Germany - Creating a livable city, 2022.
3. IKEA, Designing for a circular future, IKEA Global, 2023b.
4. Technology Magazine, How Dell's AI PCs revolutionise sustainable technology, 2025a.
5. Ahmed, F., Role of green innovation, trade and energy to promote green economic growth: Evidence from South Asian economies. *Environ. Sci. Pollut. Res.*, 28, 27, 35875–35889, 2021.
6. United Nations Environment Programme, Global Framework on Chemicals – for a Planet Free of Harm (UNEP/GFC/OEWG.1/11), 2024.
7. Mitarlis, U., Azizah, U., Yonata, B., The integration of green chemistry principles in basic chemistry learning to support achievement of sustainable development goals (SDGs) through education. *J. Technol. Sci. Educ.*, 13, 1, 233–254, 2023.

8. Balaraman, H.B., Rangarajan, V., Rathnasamy, S.K., A synergetic green approach for enhanced CO2 capture in protic deep eutectic mixtures–Kinetics and Henry's solubility determination. *J. CO2 Util.*, 65, 102217, 2022.
9. Abbott, A.P., Barron, J.C., Ryder, K.S., Wilson, D., Deep eutectic solvents applications aligned to 2030 United Nations Sustainable Development Goals: A review of process intensification and sustainability. *J. Cleaner Prod.*, 1234, Article 130097, 2024.
10. Vijay Kumar, V. and Shahin, K., Artificial intelligence and machine learning for sustainable manufacturing: Current trends and future prospects. *Intell. Sustainable Manuf.*, 2, 1, 10002, 2025.
11. Singh, R., Bhardwaj, N., Alam, P., Green nanoemulsions: Sustainable synthesis, applications, and future perspectives in food, cosmetic, and pharmaceutical industries. *J. Nanotechnol. Green Chem.*, 15, 2, 145–170, 2025.
12. Kumar, V. and Sharma, A., Cutting-edge eco-conscious technologies for glycan separation: A review of biocatalytic, membrane, chromatographic, and green solvent techniques. *Green Chem. Lett. Rev.*, 17, 4, 412–430, 2024.
13. Muniasamy, R., Venkatachalam, P., Rangarajan, V., Samal, S., Rathnasamy, S., A comprehensive perspective on sustainable bioprocessing through extractive fermentation: Challenges and prospects. *Rev. Environ. Sci. Bio/Technol.*, 22, 3, 715–737, 2023.
14. Li, X., Wang, Y., Zhang, J., Advances in deep eutectic electrolytes for sustainable energy storage: Properties, applications, and future prospects. *J. Energy Chem.*, 77, 200–215, 2024.
15. Guinée, J.B., Gorrée, M., Heijungs, R., Huppes, G., Kleijn, R., de Koning, A., van der Ven, B.L., *Handbook on life cycle assessment: Operational guide to the ISO standards*, Springer Science & Business Media, 2002.
16. Pandey, A. and Sharma, S., Techno-economic analysis and sustainability assessment of process intensification in bioactive extraction, in: *Green Technologies for Sustainable Environmental Development*, pp. 45–68, 2023.

Index

Printed and bound by CPI Group (UK) Ltd, Croydon, CR0 4YY
19/09/2026
14965093-0001